T0211818

After Childhood

This book offers a new approach for theorising and undertaking childhood research. It combines insights from childhood and generational studies with object-oriented ontologies, new materialisms, critical race and gender theories to address a range of key, intractable challenges facing children and young people.

Bringing together traditional social-scientific research methods with techniques from digital media studies, archaeology, environmental nanoscience and the visual arts, *After Childhood: Re-thinking Environment, Materiality and Media in Children's Lives* presents a way of doing childhood research that sees children move in and out of focus. In doing so, children and their experiences are not completely displaced; rather, new perspectives on concerns facing children around the world, which dominant approaches to childhood studies have not yet fully addressed, are unravelled. The book draws on the author's detailed case studies from his research in historical and geographical contexts. Examples range from British children's engagement with plastics, energy and other matter to the positioning of diverse Brazilian young people in environmental and resource challenges, and from archaeological evidence about childhoods in the United States and Europe to the global circulation of children's toys through digital media.

The book will appeal to human geographers, sociologists, anthropologists, education studies scholars and others working in the interdisciplinary field of childhood studies, as well as to anyone looking for a range of novel, interdisciplinary frames for thinking about childhood.

Peter Kraftl is Professor of Human Geography at the University of Birmingham, UK. His research focuses on geographies of children and young people, and geographies of education. He has published eight books and over 50 journal articles, including *Children Living in Sustainable Built Environments* (Routledge, 2017).

Routledge Spaces of Childhood and Youth Series
Edited by Peter Kraftl and John Horton

The *Routledge Spaces of Childhood and Youth Series* provides a forum for original, interdisciplinary and cutting edge research to explore the lives of children and young people across the social sciences and humanities. Reflecting contemporary interest in spatial processes and metaphors across several disciplines, titles within the series explore a range of ways in which concepts such as space, place, spatiality, geographical scale, movement/mobilities, networks and flows may be deployed in childhood and youth scholarship. This series provides a forum for new theoretical, empirical and methodological perspectives and ground-breaking research that reflects the wealth of research currently being undertaken. Proposals that are cross-disciplinary, comparative and/or use mixed or creative methods are particularly welcomed, as are proposals that offer critical perspectives on the role of spatial theory in understanding children and young people's lives. The series is aimed at upper-level undergraduates, research students and academics, appealing to geographers as well as the broader social sciences, arts and humanities.

Young People, Rights and Place
Erasure, Neoliberal Politics and Postchild Ethics
Stuart C. Aitken

The Common Worlds of Children and Animals
Relational Ethics for Entangled Lives
Affrica Taylor and Veronica Pacini-Ketchabaw

Intersectionality and Difference in Childhood and Youth
Global Perspectives
Edited by Nadia von Benzon and Catherine Wilkinson

After Childhood
Re-thinking Environment, Materiality and Media in Children's Lives
Peter Kraftl

For more information about this series, please visit: www.routledge.com/
Routledge-Spaces-of-Childhood-and-Youth-Series/book-series/RSCYS

After Childhood

Re-thinking Environment, Materiality
and Media in Children's Lives

Peter Kraftl

Routledge
Taylor & Francis Group

LONDON AND NEW YORK

First published 2020
by Routledge
2 Park Square, Milton Park, Abingdon, Oxon OX14 4RN
605 Third Avenue, New York, NY 10017

First issued in paperback 2021

Routledge is an imprint of the Taylor & Francis Group, an informa business

British Library Cataloguing-in-Publication Data
A catalogue record for this book is available from the British Library

Library of Congress Cataloging-in-Publication Data
A catalog record for this book has been requested

Typeset in Times New Roman
by Apex CoVantage, LLC

ISBN 13: 978-1-03-223760-2 (pbk)
ISBN 13: 978-1-138-08824-5 (hbk)

DOI: 10.4324/9781315110011

For Juliet, Emily and Adam

**This book is dedicated to the memory of
Dr David Matthews, 1980–2019**

Contents

Acknowledgements

This book represents the culmination of over ten years' worth of research projects and 20 years' worth of thought, reflection and consideration. Whilst I take responsibility for the final form this book takes, almost all of the work that has gone into it has been collaborative in some sense. This book simply could not have happened without the support of many people and organisations, to whom I am profoundly grateful.

This book would not have been possible without ongoing conversations with friends and colleagues who have offered inspiration, critique, insight, support, guidance and laughter over the past 20 years, and who either directly or indirectly shaped my work about childhood and young people. In particular, I am indebted to John Horton, Faith Tucker, Hugh Matthews, Sophie Hadfield-Hill, Pia Christensen, Alan Prout, Pyrs Gruffudd, Marcus Doel, Jenny Pickerill, Gavin Brown, Claire Jarvis, Ben Coles, Jen Dickie, Sarah Mills, Peter Kelly, Anna Sparrman, Veronica Pacini-Ketchabaw, Mindy Blaise and Affrica Taylor. Thank you also to the many other fantastic colleagues at the universities of Swansea, Northampton, Leicester and Birmingham, and elsewhere, who are too numerous to name here, but whose support for my work at conferences, at workshops and in informal conversations has been a key part of bringing this book into fruition. This book is dedicated to the memory of Dr David Matthews, my best friend during my PhD studies and without whom this book and much else would not have been possible – thank you, Dave.

The (Re)Connect the Nexus research project – funded by the Economic and Social Research Council (ESRC), São Paulo Research Foundation (FAPESP) and the Newton Fund – has been a key part of this book. I am grateful to the funders and also to my collaborators: John Horton, Sophie Hadfield-Hill, Ben Coles, Cristiana Zara, Catherine Walker, Joe Hall, Jose Antonia Perrella Balestieri and many other colleagues in the United Kingdom and Brazil. A sincere thanks to all of the young people and professionals who took part in that research.

My Plastic Childhoods Research Fellowship – funded by the Leverhulme Trust – has been central to the completion of this book. The funding, for which I am indebted, afforded not only the time but the resources to undertake the social media analyses, systematic reviews and environmental nanoscientific techniques discussed in this book. Thank you also to my collaborators: Andrew Jones, Ruth

Till and all the students at the University of Birmingham School, Sophie Hadfield-Hill, Iseult Lynch, Andy Chetwynd, Amy Walker, Polly Jarman and Alice Menzel. Thanks also to the School of Geography, Earth and Environmental Sciences at the University of Birmingham for allowing me to take research leave to work on the Fellowship and to colleagues who covered my duties during this time.

The Climate Action Network, funded by the Social Sciences and Humanities Research Council of Canada (SSHRC), provided an opportunity to undertake some novel, experimental research with organisations working with children and young people who – for a variety of reasons – seldom get the chance to take part in academic research. I am grateful to Veronica Pacini-Ketchabaw for her leadership, creativity and inspiration and for including me in the network and to the many other members of the network and the Common Worlds Research Collective for ongoing conversations about childhoods, matter and much else besides. Thank you to the St Paul's Trust – and especially the teachers, children and young people at the nursery, playscheme and secondary school for taking part so willingly in co-developing the collaborator activities. And last, but far from least, a massive thanks to Arooj Khan, for her brilliant leadership of the collaboratory activities, for some of the photographs used in this book and for critical, inspiring conversations about childhood, youth, race and energy.

Some material in Chapter 4 is taken from the EPSRC/FAPESP-funded Rehabiting the City project. My thanks to the funders and to Lauren Andres, Lucelia Rodrigues, Joana Carla Soares Goncalves, Nicole Porter, Renata Tubelo and the rest of the team.

Much of the material on energy is based on a pump-priming project, funded by the University of Birmingham's Institute for Global Innovation. My thanks to them for supporting the pilot research and systematic reviews and to Jen Dickie, Jiska de Groot and Polly Jarman for working with me to develop a programme of work around youth and energy.

Thank you to Faye Leerink at Routledge, for her ongoing support for the series of which this book is a part (and for which I am a series editor, with John Horton) – the series is going from strength to strength as a result of her vision and leadership, and I am proud that this book is a part of that series. Thank you also to the editorial assistants and production staff at Routledge, and especially Nonita Saha.

I owe a huge debt of gratitude to Pauliina Rautio and Michael Gallagher, who kindly, generously and at short notice provided some fantastic and thoughtful comments on the final draft of this book. As always, any errors, omissions, inaccuracies or failings are my own.

Finally, this book would not have been possible without my family: Juliet, Emily, Adam, Mum, Dad, Mart, Philippa, Samuel, Leo and Lyn. Thank you for your support and for putting up with me not always being completely 'present' when I was trying to write this book.

Figures and tables

Figures

Tables

1 Introduction
Thinking and doing *after childhood*

São Paulo

In 2011, the World Bank published a *Case Study Overview* reporting on the challenges facing the urban poor in terms of climate change and disaster risk. The city is experiencing a changing climate with heavier rainfall events and higher maximum temperatures alongside temporary decreases in air humidity that are, somewhat paradoxically, leaving some areas desertified. Nearly one million homes are at risk of flooding or landslides, especially in the *favelas* (poorer, informal settlements). In conjunction with a Mayor's Task Force from the city, the report highlights a number of intersecting challenges facing these communities (World Bank, 2011). Vulnerable groups often lack the resources to deal with the changing climate and, especially, the risks of flooding and landslides that are most acute on the steep hillsides where many homes are situated. Indeed, it was estimated in 2011 that around 5% of such areas were at severe risk of landslides in the subsequent 12 months. Meanwhile, in these impoverished communities, access to basic services such as electricity, water and sanitation is patchy, if present at all. What infrastructure does exist is hugely vulnerable to extreme weather events and other disasters. This includes public transport, which, in combination with the irregularity of electricity supplies, means that, especially at night, vulnerable residents feel unsafe. As a result – and because in some areas gangs control access to energy and other services – it can be difficult for children to travel to work or school, further compounding the social and economic marginalisation faced by families (especially because these areas host the highest proportions of residents aged under 19 in the city and the greatest number of child-headed households). Finally, the report highlights that children (along with the elderly) are particularly vulnerable to these compound problems – for instance, to drowning or physical injury during flooding or landslides, to going hungry or to missing out on education.

Denmark

In August 2019, the *Guardian* newspaper reported on a study demonstrating the links between growing up with air pollution and a range of mental health issues – including schizophrenia. The piece references two studies (in fact comparing the United States and Denmark) and found that the emergence of such issues in later life could not be explained by genetics alone (Khan et al., 2019; Chang et al., 2019). Comparing (somewhat bizarrely, on the face of it) insurance data, the authors found that rates of bipolar disorder were 27% higher in those counties with the poorest air quality than those with the best. Moreover, given that individuals in lower socio-economic groups are less likely to have insurance, and more likely to live in areas with poorer air quality, they suggested that the prevalence of bipolar disorder and other mental health issues was likely to be even higher amongst those populations. There are a number of possible physiological reasons for why air pollution might cause mental health issues – alongside a range of more established health problems. However, the *Guardian* (2019) also notes that, in addition to research with those in lower-income groups, further research is required to explore how air pollution intersects with a range of other potential drivers – including genetics, family circumstances, bullying and (in)formal sources of support.

Thinking and doing, *after* childhood

This book seeks to set out a range of ways for thinking about children, and doing childhood research, *after* childhood. Taken as a whole, the book offers concepts, methods and modes of narration for scholars of children and childhood from a range of disciplines, but especially the social sciences. Without wishing to claim that it constitutes a 'radical' departure from the conventions of childhood studies – not least because it draws to an extent on already existing literatures in what is a large, diverse, cross-disciplinary field – the book nevertheless puts forward and exemplifies a range of novel frames that might feel challenging, uncomfortable but hopefully exciting and generative for childhood scholars. In one sense, the book's principal reason for existing is to attempt to push even further the notions of 'child' and 'childhood' than even some of the most recent new materialist, post-humanist literatures have done. On the other hand, and despite my attachment and commitment to the field, it is born of a sense of frustration with some of those literatures, and with childhood studies more generally. Whilst social-scientific studies of childhood have achieved much in the past few decades, there is still much to do: not least in developing theories, methods and forms of interdisciplinary collaboration that might broach the kinds of challenges – and the injustices, illnesses and *traumas* associated with them – so briefly indicated by the two cases that began this chapter.

The examples from São Paulo and Denmark are intended to be indicative of two key problems, which this book seeks to broach. On the one hand, they nod to the complex, knotty, intractable challenges that affect particular communities in particular places, wherein it is not particularly clear where one would even begin in attempting to address them. In São Paulo's *favelas*, although singled out as a key cause, climate change is not the only driver of poverty, or lack of safety, or vulnerability: the challenge is multifaceted, encompassing urban immigration, education, gang cultures, the provision of infrastructures in informal urban settings, the interactions between slopes and ad hoc housing structures, and far more besides. In Denmark (and the United States), the drivers of mental health issues are complex, taking in genetics, the interactions between environmental pollutants, the respiratory tract and the brain, transport planning and reliance on the car, socio-economic inequalities, bullying and, again, far more besides.

On the other hand, this book does not purport to 'solve' these intractable, complex challenges. Yet it seeks to look again at how *children* are positioned within them – as I ask in Chapter 3: *where are* children located, precisely, in complex nexuses of resources, institutions and more-than-human interactions? Moreover, it seeks to construct an argument that one of the best ways to understand, analyse and respond to these intractable challenges – and children's position within them – is to recognise that, sometimes, we cannot *start* with children (or childhoods) at all. For, in a very simplistic sense, the two examples that began this chapter are not just about children or childhoods. As I argue, this means, to differing extents, a loosening of control: perhaps decentring childhoods (Spyrou, 2017); perhaps allowing childhoods to move out of *focus*; perhaps engaging (after Tsing, 2015) in arts of *(not) noticing*; and, ethically and politically, as I argue throughout the book and especially in conclusion, holding apparently contradictory conditions such as *silliness* and *trauma* in productive tension. For conceptual, methodological, political and ethical reasons, as I argue, this does not mean going *beyond* childhood. Rather, as I demonstrate in Chapter 9, it means engaging with the ethical and political possibilities (*both* speculative-playful-*silly and* marginalising-violent-*traumatic*) constituted by thinking and doing, *after childhood.*

At this early stage, I want to be clear that I see the idea of thinking and doing, after childhood, as at once a starting point for a series of (sometimes perhaps esoteric) thought experiments *and* a way of responding, differently, to the very real challenges experienced by *diverse* children and young people, in different places, in different ways, around the world – without making any claims to universality. Thus, before moving on, it is helpful to clarify just some of the ways in which I seek to understand and theorise the term 'after childhood', which are in turn woven through this book.

Firstly, and most straightforwardly, the term 'after' denotes a series of *temporal* logics. Although diverse, the notion of childhood is – and arguably always has been – characterised by a temporal logic that places children in generational orderings (Alanen and Mayall, 2001; Punch, 2002). Herein, children have often been viewed as future adults – as *becomings* – upon whose shoulders rest a range of hopes and fears (Uprichard, 2008; Evans, 2010). In these logics, even if any given children's childhoods come 'after' those of the previous generation's, childhoods are most frequently understood to come 'before' adulthoods and are defined by what adulthood is not (maturity, physical size, status, etc.).

These debates will be familiar to many childhood studies scholars; thus, throughout the book I ask what work the term 'after' might do in twisting some of these apparently simple, modernist temporalities. Certainly, several childhood studies scholars have sought to query and/or queer the teleologies inherent in the age-old 'beings-becomings' debate. Yet the possibilities for questioning these linear generational orderings and temporalities have not been exhausted. Thus, for instance, via concepts like 'hyperobjects' (Morton, 2013), I examine the phasing of material stuff and processes – like plastics – into human lives over a much longer temporal frame. This is in turn key to my concept of 'infra-generations' (Chapter 6), although I take a very different stance from those (like Postman, 1985) who write on the social status and experience of childhood over time. Mine is not an argument that childhoods are either changing or disappearing, although both might be happening, in different ways, in different times and places, as the final provocation in Chapter 9 makes clear. Rather (for example), drawing on archaeologies of childhood, and starting with found *objects* – which immediately bring to mind a much longer chronology than the usual frames of childhood studies – the term 'after' can be understood in a different light. Some of the 'toys' uncovered in Chapter 5 were played with by children *after* they had been thrown away by adults, reversing the usual ways of things (i.e. where adulthood comes 'after' childhood). The toys, bones and other objects found in 'cuts' were found *after* the death of the subjects who played with or were constituted by those objects, requiring rather different modes for deducing and analysing 'childhoods' than notions of voice or agency. Indeed, as I come to below, this latter observation, when taken to its most logical conclusion, imagines a world without humans (and, hence, without children). Using the term 'after' as a springboard, throughout the book I toy with various temporalities and chronologies – fast and slow, big and small, linear and, especially, non-linear.

Secondly, given its complex etymology in English, the term 'after' also implies something perhaps subtly, perhaps more radically different than the prefix 'post-', which extends beyond temporality. As a preposition it denotes repetition and circularity ('time after time') and attendant responsibility ('clean up after yourself') rather than only consequence and linear chronology. It signals that which comes behind ('shut the door *after* you leave') and is, like my deployment of the related prefix 'infra-' in Chapter 6, somewhat modest ('play often comes *after* education and health in priorities for policies') yet also a pursuit for something more ('chasing *after* a dream'). 'After' implies allusion or imitation ('*after* the style of this musician'): an impulse that finds expression throughout this book in the deployment of speculative fabulation (Haraway, 2011), listing (Latour, 2005), metaphor (Bogost, 2012) and a call for 'radical', experimental forms of interdisciplinary childhood studies (see, especially, Chapter 7). And, finally, it denotes a consistent double logic (sometimes referred to as a 'pull focus') throughout this book of the 'after-all': after all, the world goes on and will exceed childhoods (and all humans); but, after all, children *and* childhoods matter, in that ongoingness, in ways that require much more thought and action. Ultimately, unlike some brands of post-(humanist) thinking, mine is *never* a position that endorses a way of thinking and doing *without* humans (or without children) and the things that concern us (and them). That would be unsustainable in a book about childhood. Thus, in certain places I do stray far from children and childhoods and sometimes speculate about worlds without humans. However, my commitment to the 'pull focus' means that children and childhoods – and sometimes fairly conventional ways to understand their voice, agency or mobilities – are woven (back) into the narrative, even if in oblique or apparently belated styles (Chapter 8 being a case in point).

Thirdly, the idea of looking 'after' evokes decades of 'post-' thinking: of attempts to go 'beyond' (Bryant, 2014; Jackson, 2015). The prefix 'post-' does not just imply that which follows chronologically, but of attempts to complicate, exceed, subvert or question: to go beyond the – modern, the – colonial, the – structural, the – truth. Indeed, as discussed in Chapter 2, this book aligns with what some have termed a 'new wave' (Ryan, 2012) of childhood studies or, more explicitly, theories of the 'posthuman child' (Murris, 2016) or 'post-child' (Aitken, 2018) that are slowly gaining traction.

However, as implied above, the prefix 'post-' brings forth a range of connotations that are, admittedly, semantic – but which go beyond mere wordplay. On the one hand, it is loaded: it now implies a range of modes of theorising that, however generative and progressive, have *also* become associated with a continual search for newness, with a sometimes white, and/or masculine and/or middle-class and/ or European impulse and aesthetic. As I highlight in the next chapter, in *some* forms, the impulse to go 'beyond' can – perhaps unintentionally – have the effect of reinforcing privilege and marginalisation. Although writing from the perspective of yet another white, middle-class, European male, I find the term 'after' – or rather I *want* the term 'after' – to feel somewhat more modest, humble and cognisant of, if not a challenge to, these modes of marginalisation. Thus, limited by both my positionality and skill, I attempt throughout the book to engage meaningfully and consistently with feminist, queer and critical race theorisations of difference and matter that sit alongside and in tension with certain forms of 'post-' child thinking (including deployments of actor-network theory (ANT) and new materialist theory).

On the other hand, the idea of the 'post' is both too much and not enough, all at once. When coupled with childhood, the effects of post-child thinking (especially in the most popular, feminist/new materialist/post-human brands) are to attempt to 'decentre' children and childhoods in favour of non-human animals and materials (Spyrou, 2017). This decentring is not enough: the very language implies an anthropocentrism that *starts-with* the human and with analyses and politics that "still preserve humanity as a primary actor" (Bryant, 2014: 7). It also tends to decentre childhoods for certain kinds of non-humans – animals and neatly bounded objects like toys – such that other messier, less tangible, less object- (or subject-)like stuff (like plastics, digital media and energy) is effaced. And this decentring is too much: at least as things stood when I wrote this book, it enabled a range of more generous stories but, bizarrely, seldom lived up to the promise that post-human approaches could have in broaching some of the intractable challenges facing *children* (such as the two that preface this chapter). Not only am I troubled by the prefix 'post-' and the concept of 'decentring' politically, then, but, as I argue throughout this book, that decentring is not (yet) sophisticated enough to grapple with children's positioning in resource nexuses (Chapter 3), their visualisation and circulation in digital media (Chapter 5), or their knowledge and experiences of energy (Chapter 7). For me, this requires a double manoeuvre that both *exceeds* and is more *modest* than the notion of 'decentring': what I term a 'pull focus' or, following Tsing (2015), "arts of (not) noticing".

Fourthly, the term 'after' brings with it a specific meaning when it comes to matter and materiality: the very *stuff* of new materialist or post-human childhood studies. Alongside the generative approaches of feminist, queer and critical race theorists, this particular interpretation of thinking and doing 'after' requires a foray into speculative-realist and object-oriented ontologies (OOO).[1] These theories are introduced in Chapters 2 and 3 – but perhaps at this point three key contentions of OOO are important to note: that the world can exist after human finitude (i.e. the sense that the world has existed, and can exist again, without humans); that 'objects' can and do exist and operate outside of relationships with humans (i.e. a landslide can happen on a remote mountain without humans witnessing it, and it will have happened whether or not we discover the aftermath); and that the relations between objects do not exhaust their very being (i.e. a human could look at a mountain from many angles but never 'see' the whole mountain, insides and all, without totally dismantling it). OOO theorists are sometimes known as anti-correlationists because they push against the widely held view – central to both modernism and post-modernism – that the world only exists as humans perceive, understand and represent it (i.e. 'reality' only exists insofar as it correlates with human perception). For many contemporary thinkers – not just those falling under an OOO banner – this is a form of human exceptionalism that is arrogant and dangerous, particularly when it comes to the question of who is able to make claims about the world and whose interests are served in doing so. Importantly, then, as I do with feminist, queer and critical race theories, I draw on OOO as a provocation, and in combination with a range of other theoretical and empirical analyses, in order to speculate about media, resources, generations, energy and more besides, as they 'cut' across and matter to diverse childhoods (see Chapter 3).

One vital aspect of this provocation is a need to engage in *methods* that challenge – perhaps radically – the doxa of childhood studies. In particular, I discuss throughout this book various forms of interdisciplinary and transdisciplinary collaboration that involve techniques and perspectives rarely, if ever, deployed in the field – including work with artists, architects, engineers and environmental nanoscientists. As I argue forcefully and repeatedly, I *never* see (for instance) nanoscience as *the* or even *a* solution for the kinds of intractable challenges in which children are entangled. Rather, I promote and attempt to exemplify forms of humble, creative, speculative, generous but nonetheless robust *collaboration* in which the 'conventional' social-scientific and phenomenological commitments of childhood studies still have an important part to play.

Finally, then, there remains at least one question: how to respond to these overlapping understandings of and provocations posed by the term 'after', in a book about *childhood*? Key here is that I will not take speculative realist, OOO or any other positions as 'read': this book is not a wholehearted endorsement of those philosophies nor of the non-representational and new materialist approaches with which I have been working for many years (for an introduction, see Chapter 2). Indeed, some of the critiques of those kinds of approaches by critical race and queer philosophers – alongside the inclusion of a range of other theorists included in this book – are testament to my commitment to using the term 'after' to pose questions about substantive issues and challenges in which children are (sometimes, in some places) involved in ways that are not faithful to or exemplary of any one philosophical position. Yet I take from OOO, especially, an injunction to *lose control* a little: to remain absolutely concerned with children but to let them slip from view – to move out of focus. Perhaps, ultimately, my reason for doing so is *anti*-anti-correlationist because I argue that in *not* noticing, in *not* paying attention to children at some junctures, it might be possible to undertake even more powerful analyses of that which matters to children (and, as I argue in Chapter 5, those aspects of, for instance, contemporary digital cultures that have gone underscrutinised as a result). I also toy with OOO – literally – by asking in Chapter 6 how to deal with 'what lies beneath'; when one cuts or scratches the surface of a toy, or the earth itself, what that means for the putatively 'withdrawn' properties of an object.

Critically, however, I seek to enact a double manoeuvre that extends beyond both the approaches and the languages of contemporary childhood studies by asking what it might be like to *think and do after childhood*; but that takes from OOO and elsewhere a sense of humility. Contrary to what we often think and feel (at least in Western cultures), neither humans nor children are the centre of our world. This is not merely a matter of philosophical experiment, or even of the new methodologies and conceptual languages that I develop throughout the book – although all are responses to the questions I have raised in this section about that which the term 'after' might provoke for childhood studies. Rather, it is a vital acknowledgement that humans are hurtling towards a time after finitude, after the death of our species as we know it; that future remnants of that time already exist (in, for instance, the environmental degradation and human illnesses already being caused by plastics and oil); and that, if we just let children and childhoods out of focus, in a range of ways, it becomes paradoxically *clearer* that children and childhoods are, actually, co-implicated in those 'after-lives' in all kinds of troubling ways, which exceed a simple recognition of children being 'the next generation'. Thus, as I argue in Chapter 2, childhood studies requires more than a move 'beyond' bio-social dualism, or an attunement to 'entanglements' and forms of 'decentring', but a more fundamental shift towards different languages that are somehow more radically out of control whilst remaining more modestly committed to the subject position named 'child'.

Methods: researching, after childhood

Although drawing on literatures and examples from a range of sources, the backbone of this book is a series of major, funded, original research projects about children and young people, their everyday lives and their environments. Whilst not initially conceived as such, these projects have several overlapping themes that mean that to all intents and purposes they constitute a programme of research about climate change, resources (especially energy), media and materialities. Whether singly or in combination, an original analysis of empirical materials from these projects informs all of the chapters in this book. Therefore, to aid reference and avoid repetition, I introduce the projects and their methodologies briefly below, alongside websites and/or relevant publications that provide further information.

The (Re)Connect the Nexus project (2016–2018) was a major 20-month research collaboration that aimed to examine children and young people's (aged 10–24) experiences of and learning about the food–water–energy nexus in Brazil. The project involved a collaboration (established in 2014) between British-based human geographers and engineers and education scholars at São Paulo State University (UNESP) in Brazil. It sought to generate interdisciplinary insights into, and solutions for, the opportunities and threats faced by children and young people in relation to the environment and environmental resources (characterised as 'resource-power' in Chapter 2). In sum, we engaged with nearly 4,000 children and young people, living in the Metropolitan Region of Paraiba do Sul River Basin and São Paulo State North Shore. The region – which sits between the major cities of São Paulo and Rio de Janeiro – was chosen because it is socially, economically and geographically diverse and faces a number of interrelating nexus 'challenges' (somewhat akin to the 'intractable' challenges discussed earlier). Home to 2.1 million people, it is on the face of it one of the richest areas of Brazil, crossed by the Via Dutra, the main highway between São Paulo and Rio de Janeiro. Thus, it is home to tourism (in the mountainous area around Campos do Jordao); to government, education and technology industries; and to a vast patchwork of agricultural landscapes – from valley-floor rice fields, to sugar cane farms producing both sugar for food and for bioethanol, to eucalyptus plantations from which paper is derived. However, upon digging deeper, the region is beset by inequalities – from informal and impoverished settlements in smaller cities that pale in comparison in scale with those in the major cities but that nevertheless suffer from many of the same problems to rural poverty in the agricultural regions and along the State's North Shore (Litoral Norte). The challenges faced are therefore also diverse and intensify with climatic changes – soil degradation, landslides, flooding and, paradoxically, drought, all play a role in and intersect with children and young people's everyday experiences of food, water and energy.

The project itself included four strands: qualitative research with 48 young people; semi-structured interviews with 64 key professionals from relevant sectors; a large-scale, in-depth survey with over 3,700 young people living in the region; and a global video competition (for more details, see Kraftl et al., 2019; for a list of publications on the second, third and fourth strands, see www.food waterenergynexus.com). The first (reported on in detail in this book) was a programme of in-depth research with 48 young people, aged 10–24. Young people were recruited through schools, workplaces, community groups and NGOs, and we ensured that – within the bounds of qualitative research – they were broadly representative in terms of gender and socio-economic status (using established Brazilian classifications to determine the latter). Research with each young person was intensive: an initial interview, taking between one and two hours, which explored their experiences of different elements of food, water and energy; use of a bespoke nexus smartphone app that asked young people to take geotagged photographs of and comment on aspects of food, water and energy that they encountered during their everyday lives; a follow-up interview (often multiple interviews) lasting sometimes several hours that discussed outputs from the app – most importantly in the production of an individual 'visual web' for each young person (see, for instance, Figure 3.4). During that latter stage, young people worked collaboratively with the researchers to choose photographs, place them onto a large sheet of paper and effectively create their own personal 'nexus' by drawing on connections and making other observations and illustrations as the conversations progressed. Most of the interviews were conducted in Portuguese, were audio recorded and were translated and fully transcribed before being subject to a detailed thematic analysis involving the whole team.

The Climate Action Network (2018–2020) involved an interdisciplinary collaboration across three countries (Canada, Australia and the United Kingdom) and was based within the feminist, new materialist, decolonising principles of the Common Worlds Research Collective (https://commonworlds.net/). As a whole, the network sought to imagine, co-construct and experiment with novel, alternative, perhaps 'radical' forms of environmental pedagogy. The focus was generally on younger children, although the Birmingham element (which I led) included work with children aged between 2 and 16. The project as a whole worked through the principle of developing a number of collaboratories, involving researchers, child-/youth-professionals (mainly teachers and day-care providers) in different sites in each country, which were to both provide 'data' for the project and also outlast the research. Each collaboratory was themed: the Birmingham theme – given the city's rich industrial heritage (discussed in depth in Chapter 7) – was 'energy'. However, rather than focus only on conventional forms of energy education (i.e. technologies for producing or behaviours for saving energy), we adopted a very broad definition, including forms of embodied, emotional and material energies. Other themes included plastics, rain, trees, flooding and waste (see http://climateactionchildhood.net/).

The collaboratories all took different forms, but the key driver was an elision of collaboration with experimentation (the 'laboratory' element of the neologism 'collaboratory'). Each university team engaged in a long-term relationship with a key educational site. In Birmingham, we worked with St Paul's Community Development Trust (www.stpaulstrust.org.uk/). The Trust was established in 1973, in Balsall Heath, an area of largely terraced housing, fairly near the city centre, that initially housed workers in Birmingham's large industrial economy in the late nineteenth and early twentieth centuries. Today, the residents of Balsall Heath are rather different. As a result of other forms of energy – Birmingham being a key locus for immigration during the mid- and late-twentieth century – more than 50% of the population are British South Asian, and particularly of Pakistani descent. The area is also very young in demographic terms – more than 45% are aged under 25, which is far higher than the national average. And Balsall Heath also experiences fairly high levels of socio-economic deprivation – a key indicator being that unemployment (in 2019) sat at over twice the UK average, at 18%.

As a community-based and voluntarily run organisation, the Trust works and has worked to provide educational services and support for the residents of Balsall Heath. At the time of writing, it ran several nurseries and day-care centres, a secondary school for young people (largely teenage boys) who had been excluded from mainstream schools because of behavioural, emotional or social differences (BESD), a city farm and a canal boat. Aside from the secondary school, which took referrals from across the city, the Trust's sites were intended to be centred around the needs of local residents. We worked with the Trust's management and with professionals at each site – as well as with children themselves – to develop a series of activities on the broad theme of 'energy'. These included several 'energy walks' around Balsall Heath, which encouraged children and young people to think about the energy-related challenges facing the area; artistic activities (mainly with younger children), involving painting, throwing, sculpting and making, which sought to explore magnetic, kinaesthetic and gravitational energies; workshops at a local science museum and the use of a number of kits and resources to make alternative energy 'technologies', such as wind turbines and solar panels; visits to other key sites across the city, including an exemplar ecohouse and a geological museum; opportunities for free play, imagination and the generation of "speculative fabulations" (Haraway, 2011; see Chapter 7); and, more formal, classroom-based sessions, including opportunities for reflection, questioning and feedback. Although working in somewhat of a post-qualitative frame and thus questioning the status of 'data' as such, we collected materials and took detailed ethnographic observations of and reflections from our work over the period of a year, even though our engagement with the Trust exceed that period. In all, we worked intensively with 49 children during that time (19 from the play-schemes and nurseries, 30 from the secondary school).

The (Re)Inhabiting the City project was a year-long (2018) collaboration between human geographers, architects and planners at the University of São Paulo (USP, Brazil) and the universities of Nottingham and Birmingham (UK). The main aim was to explore the opportunities and challenges of 'temporary urban interventions' for dealing with perceived vacancy and dereliction in the urban core of São Paulo and cities like it. As Chapter 4 outlines in more detail, we worked in the neighbourhoods of Luz and Santa Ifgênia. Both are close to the city centre and well connected through the city's bus and subway networks; but both suffer from high levels of socio-economic deprivation, drug use, informal (and sometimes illegal) squatting and physical dilapidation, meaning that many of the twentieth-century apartment blocks characteristic of the areas (Figure 4.3) are abandoned and in a state of disrepair. Drawing on the work of Andres (2012) and Madanipour (2018), we critically analysed the potential for temporary uses – like a container theatre, benches, picnic tables, table tennis tables, play equipment – to 'enliven' these spaces.

Thus, we were interested in the capacity for more minor interventions than formal attempts at regeneration or gentrification – that worked from the bottom up with support from the city municipalities (the *Prefeitura de São Paulo*) to encourage urban residents to be present, to interact and to use public spaces.

Although the team did not directly work with children, matters concerning children and young people cut across the project in various ways – from observations of children's play in what was a challenging place to grow up, to the installation of play equipment, to the wider demographic context that these are (even in Brazilian terms) demographically 'young' neighbourhoods. Methodologically, we worked together to develop observational methods and post-occupancy analyses that drew on established methods from our respective disciplines: ethnographic observations and note-taking; measurements of light intensity, sound levels, heat and humidity; quantifications of pedestrian and other traffic; land-use mapping; sketching and photography. These methods were developed over a six-month period and focused upon particular carefully chosen sites in Luz and Santa Ifgênia (both with and without temporary interventions).

The Plastic Childhoods project (2018–2020) forms a major component of the empirical research and the conceptual work for this book. The project aimed to explore the many ways in which childhoods and plastics are 'entangled' in both historical and contemporary contexts. Although taking the form of an individual fellowship, the funding for the project enabled collaboration with artists, digital media analysts, environmental nanoscientists and schoolteachers, as well as with a core group of 13 young people at a secondary school in Birmingham. The 18-month project began shortly after the airing of the final episode of David Attenborough's flagship *Blue Planet II* series on the BBC, which drew attention to the problem of plastics in the world's oceans. Although framed by serious and multiple environmental and health concerns about plastics (see Chapter 9), the project sought to take a step back: to ask what were and are the ways in which children's lives are implicated with plastics and to ask how these might be valued or judged in ways that are not necessarily all negative. The project comprised three components, two of which (the digital media analyses and biosampling), in their particular form and scope, had, at the time of the research, never before been attempted by childhood studies scholars and were thus, to my knowledge, entirely novel.

The digital media analyses (discussed in more detail in Chapter 5) involved interrogating posts on Twitter and a major online selling platform.[2] For Twitter, an Application Programming Interface (API) was used to 'scrape' for tweets related to childhood and/or plastics and/or the environment. In collaboration with Andrew, the digital methods specialist, I developed a lexicon of search terms that we refined as tweets were scraped in the first month of data collection, including a number of hashtags relating to relevant recent events (such as the #climatestrikes). The TwitteR R package was used to perform the scrape and then convert the data to .csv files. The files contained the text of the tweet, the user ID (not included in the book to preserve anonymity), date/time information and the number of retweets and favourites. Tweets were scraped between November 2018 and March 2019, yielding over 1.3 million tweets. From there, we analysed the data in a number of ways. In this book I focus on hashtag networks (visualisations of tweets and retweets, as in Figure 5.1); tweet density (a graphical depiction showing the intensity of tweets over time, as in Figure 6.3); and – acknowledging the challenges of doing so – a sentiment analysis of 'positive' and 'negative' tweets using the get_nrc_sentiment function from the syuzhet R package (see Chapter 5). In addition, after a detailed read through these tweets, the text of particular tweets was taken as indicative of particular debates, sentiments or material things.

For the online selling platform – and for what we termed 'online archaeologies' – a similar API was used to search for particular kinds of objects put up for sale. We concentrated very deliberately on a subset of objects that were stereotypically redolent of modern, Western (especially British) childhoods and with gendered norms associated with them – such as toy cars and dolls. An initial scrape – which yielded similar results to the Twitter scrape in terms of text, users, etc. – was then refined into a number of thematic categories, guided in part by the online selling site's interface (see discussion in Chapter 5), in part by the terms used by users in posting objects for sale and in part via our own iterative analysis of obvious themes/categories we were finding (Table 5.1 provides an overview). In total, the scrapes yielded nearly 6.7 million items. After a detailed read through subsets of entries for each category, we chose particular objects for further analysis – capturing indicative examples for each category and sub-category and taking screenshots of images and text about the items for sale.

The biosampling work was part of a programme of interdisciplinary research in a local school in Birmingham. In collaboration with environmental nanoscientists, artists and teachers, we conducted intensive, post-qualitative and cross-disciplinary research with a group of 13 secondary school children, aged 11–15. The children opted to take part in the workshops – on 'plastics' – as their chosen enrichment session for the summer term of 2019 (April–July). Of the children who took part, eight were boys and five were girls. The school recruits from diverse areas of Birmingham and, although in such a small sample it is of course not possible to be representative, students came from very different parts of the city, from diverse socio-economic backgrounds, and different ethnic groups (five were White British; the remainder were British Asian, although from families with ties to Iran, India and Pakistan).

The students took part in a programme of workshops that we co-developed with the school. Some of the workshops were fairly straightforward opportunities to learn about and discuss plastics – indeed, it transpired that in many cases, the students knew more than we did and were well versed on recent coverage about plastics (part of the reason they had chosen to take part in the project). Other workshops required different forms of engagement, however. In one, we encouraged students to engage in a more tactile, material, multisensory way with plastics by attempting to sort them into different plastics (there are six main types used in most forms of packaging) not only by sight and by technical markings but by the very feel of the plastics. In another (described in detail in Chapter 8), we worked with local artists, General Public, to create a series of 'totem poles' designed to disrupt our (and their) by-then established knowledges of and relationships with plastics. For another, we developed a bespoke Plastic Childhoods app (based on the (Re)Connect the Nexus app), which enabled students to take photos of and comment on plastics they found in the course of their everyday lives, routines and movements around the city. The app activity was followed up with an interview, lasting around an hour, which took a similar form to the 'visual web' exercise undertaken with Brazilian young people.

The final, most groundbreaking (literally, in some senses) and also most controversial part of the workshops was the use of biosampling. After a nine-month process of obtaining ethical and risk assessment clearance (about which more details in Kraftl et al., forthcoming), all of the students were trained to take samples of tap water and soil, using standard equipment used by environmental scientists for the collection of environmental samples. In addition, some (but not all) of the students consented to give samples of breath (via an EBC or Exhaled Breath Condensate instrument) and urine. All of the samples were analysed at a laboratory at the University of Birmingham using Fourier-transform infrared spectroscopy (FTIR) and inductively coupled plasma mass spectrometry (ICPMS), which are able to determine – usually in parts per billion (ppb) – the presence of 27 elements (mainly metals) as well as plastics, as they circulate around, into, through and out of children's environments and their bodies. Uniquely, as I discuss in Chapters 4 and 8, this offered a way to 'decentre' children – and to explore 'entanglements' of the bio and the social – in ways that are not only unprecedented in childhood studies scholarship but, at the time of writing, had never been systematically used and analysed by even environmental nanoscientists themselves.

Beyond the digital methods and workshops, the Plastic Childhoods project provided resources and time for a systematic review of both academic and other literatures (including news coverage) about plastics, childhood and the environment, as well as of fascinating work in archaeologies of childhood. These materials were supplemented by secondary research for the Energy Beyond Technology project. The latter provided resources for a series of systematic reviews about academic research on, and key global case studies of, children's engagements with energy – particularly beyond learning about sustainable energy technologies. Materials from these reviews are analysed throughout the book, but especially in Chapters 5 and 7.

Critically, in the Plastic Childhoods and other projects – and in accordance with Bogost (2012), whose work I discuss in depth in Chapter 8 – the intention was never, and should never be, to view engineering, or architectural, or digital media, or biosampling techniques as 'solutions', either to the questions posed about the status of childhood studies (e.g. Punch, 2019) or ongoing questions about the potential for new materialist and post-humanist methods to *really* 'decentre' childhoods (both discussed in Chapter 2). Rather, and again in line with my intention to think and do *after* childhood, they were used to carefully but significantly rephrase these questions: to ask what it might mean for children and childhoods to move in and out of focus, in arts of (not) noticing (Chapter 4). They also required engagement in deep, detailed, time-consuming (in some cases more so than actually collecting the data), sometimes difficult, but always rewarding *conversations*, across disciplines, wherein it became clear, for instance, that the biosampling results *only* made sense in light of what we *also* knew about children's everyday routines, mobilities and agency – from the app and from simply asking them. I am not suggesting that interdisciplinarity is the solution either – indeed, I develop in this book a set of further insights and tools around speculation, narrative and modes of thinking that perhaps require interdisciplinary insights to make them work but could be deployed in other (perhaps more conventional, perhaps more radical) ways. Indeed, the whole premise of thinking and doing, after childhood, is not to invent or impose a new paradigm, but it is to introduce a set of new ways of thinking and doing that I hope might inspire a range of responses and which might enable childhood studies scholars to push, even further, at the challenges invoked by non-representational, new materialist, post-humanist, OOO, critical race, queer and generational theorists.

Writing, after childhood

The structure of this book is organised around the development of a number of different conceptual, methodological and stylistic tools (and terms) for thinking and doing, after childhood. It also builds towards an ethical and political commitment to interrogating how *silliness* (and a range of related terms) and *trauma* (and a range of related terms) *interface* and operate in generative tension as I speculate on responses to the kinds of intractable challenges that began this chapter and the differential positioning of children therein. Rather than follow the singular and potentially flawed logic of 'decentring', each chapter offers a different cross-section around, through and away from childhoods. In each chapter, childhoods move in and out of focus in different ways, and in relation to different conceptual and empirical themes. And, indeed, in each chapter, I experiment – to an extent – with different modes of presentation, writing and narrative. As I argue throughout the book, and in conclusion, these different modes of (re)presentation are as important as 'high' theory to stretching ways of thinking and doing, after childhood.

Chapter 2 provides a more detailed introduction to childhood studies and, especially, to attempts over the past decade to 'decentre' children in a range of ways through forms of 'post-' thinking. Whilst a key point of inspiration for this book, it then considers a range of critiques of and ways on from that work and, in doing so, introduces feminist, queer, critical race, object-oriented and generational theories that are also important for analysing plays of difference, matter, time and space across childhoods. In doing so, I also grapple with the question of how – given my positionality as a privileged, white male – it might be possible not only to acknowledge this but to work with (for instance) critical race scholarship on materiality and toxicity in ways that could be sensitive and generative, even if always limited.

Chapter 3 offers a more detailed discussion of complex, 'intractable' challenges facing children and young people, whilst introducing the idea of arts of (not) noticing. Although using a similar term, the chapter's key argument – exploring a number of 'cuts' through childhood and through the earth – is not to any great extent connected with Barad's (2007) notion of the agential cut. Rather, my use of the term 'cut' is somewhat more literal, because in Chapter 3, and again in Chapter 6, I often start with *literal* cuts into the earth (a cross-section at an eco-park, for instance) or into childhood objects (like a damaged toy). Empirically, the chapter focuses on the uncertain, sometimes oblique, positioning of young people in Brazil in relation to the food–water–energy nexus in that country. I examine the opportunities offered by the concepts of nexus thinking and what I term 'resource-power' for connecting ineluctably multifaceted environmental processes with questions of social and environmental justice that detain many young Brazilians. As elsewhere in the book, children and young people weave in and out of the narrative – although this weaving works in different ways, with different intensities, at different points in this chapter and others. Thus, writing as a geographer, I ask a question that haunts the rest of the book: *where are children, precisely*, in attempts to think and do, after childhood?

Chapter 4 looks at matter. It is a largely conceptual intervention that introduces in more detail two object-oriented theorists whose work I have found particularly useful – but with which I retain a critical and slightly uncertain relationship. It offers an assessment of the possibilities of Timothy Morton's (2013) 'hyperobjects' and Levi Bryant's (2014) 'onto-cartographies' for thinking and doing, after childhood. Writing as a geographer, I find the spatial tropes in their work – of scale, phasing, interface – as helpful as some of the ways in which they write about how objects (or 'units' or 'machines') operate on one another. Critically, I find that, amongst the range of speculative and object-oriented thinkers, they offer some of the clearest possibilities for thinking after the human, or for conceiving post-phenomenologies, in ways that might leave space for human constructs and experiences – such as childhood. Drawing on examples taken from Brazil, the Democratic Republic of Congo and the Caribbean Sea, the key contribution of this chapter is to outline a set of key conceptual tools and languages for thinking and doing, after childhood. The chapter outlines how notions like 'local manifestations', 'unit operations', 'viscosity' and 'phasing' could offer new ways to broach the question: in a world of objects that both relate and are partly withdrawn from one another, *where are children, precisely?* Moreover, following Chapter 3, this chapter offers another way to consider how children are (not) entangled with matter – like plastics – in ways that matter, profoundly (from controversies around child labour to the harmful effects of plastics, circulating around the world's hydrological systems).

In Chapter 4, Bryant's (2014) rather particular understanding of the term 'media' is introduced. In Chapter 5, Bryant's work is used as one of a series of springboards for thinking through the ways in which children and childhoods are positioned (again, *where they are, precisely*) in relation to digital media. Since I am inspired by OOO and particularly Morton and Bryant, but also retain something of a distance from some of their perhaps more 'radical' contentions, this chapter also draws upon a range of digital media and technology theorists, some of whom have written directly on childhood and youth. Empirically and, to an extent, politically, the chapter is cognisant of but steers a different path from some of the emotive debates that surround both the representation of young people in digital media and their use thereof. I do not at all deny that, for instance, social media may have pernicious and long-lasting effects on children's (mental) health – even if recent research demonstrates some advantages for children's health if those media are used in particular ways. Rather, and unlike the vast majority of work about children and digital media, I ask a completely different question: *how* does stuff *circulate*, about childhoods? This means, again, retaining an interest in objects, although, through the use of innovative social-media-harvesting techniques developed in the Plastic Childhoods project, I also lose control of objects – at least as they are understood in the previous chapter. Instead, Chapter 5 develops three key strands that, to my knowledge, have never been explored by childhood studies scholars. First – and without forgetting classical notions of the social constructedness of childhood – it asks how childhoods *circulate* and are made *visible* in new, massive, intensifying ways through social media networks. Second, through an analysis of a popular online selling site, it considers the kinds of *interfaces* that exist between humans, technologies and particular objects as the stuff of childhoods – like toys – circulates via digital media. Herein, I develop a sub-series of terms that may prove useful for future analyses of such interfaces, which I loop back to in Chapter 9. Third, and returning to the question of visibility, it examines the implications of the *massification* of stuff – texts, videos, images, objects – for the ethics and politics of inclusion, and particularly of children and young people repeatedly marginalised on the basis of their ethnicity, disability or socio-economic positioning.

Chapter 6 returns to some of the concepts developed in Chapter 3, by setting the notions of resource-power and the 'cut' into a longer temporal frame. Developing a new term – 'infra-generations' – it asks and systematically exemplifies how more-than-human generations may be theorised. Like other chapters, a degree of *speculation* is required, which I hope is not frustrating, and is intended not to replace but offer a different kind of *robustness*. I ask whether earthly processes (like climate change) or material processes (like the biography or degradation of an object) might somehow be conceived as 'generations' – if only to afford a much clearer sense of how the temporalities of those processes accompany, produce and suture *human* generations. The chapter weaves together two 'sources': an original reading of archaeological literatures on childhood (that in turn use a dose of scientific speculation in their analyses) and a further consideration of objects found on an online selling site. In terms of a deepening of the discussion around resource-power, and anticipating arguments in Chapter 9 about marginalisation, death and *trauma*, the chapter explores the entanglement of human generations with/in a range of intractable, complex processes that *matter*. For instance, an analysis of archaeological records can bring to light the ways in which changing weather patterns, disease and migration led to the demise of the Roman Empire. Moreover, evidence from more recent sites in North America offers a stark and poignant insight into how children were positioned during the socio-economic decline of a town bypassed by the main phase of railroad construction.

Critically, archaeological *methods* provide a valuable insight into arts of (not) noticing, or what in Chapter 4 I term, following Morton (2013), the 'pull focus'. That is, the search for traces of *childhood*, specifically, is fraught with difficulties and ambiguities; yet it is sometimes by precisely *not* looking for *obvious* expressions of children's agency (in objects categorised as 'toys', for instance) that childhoods can nevertheless be found. At other points, and anticipating a point I make in the final two chapters, it is in knowing precisely how and, especially, *where* children exercise their agency (where, as decades of childhood studies scholarship has shown, children play in public spaces) that some of those ambiguities can be resolved. Meanwhile, the chapter develops in far more depth the notion of the "cut": whether in terms of the archaeological 'cut' (the midden heap or dumping ground), or what I argue are its modern counterparts in the form of online selling sites; or in terms of the damage and trauma experienced by objects and bones as they were used and divested. Thus, this chapter offers one of several moments in the book where – in provocatively asking 'what lies beneath (the surface)?' – an analysis of childhoods and the (cut-into) objects associated with them that could speak back to OOO, new materialisms and their propositions about matter.

Chapter 7 deals with the different ways in which *energy* crosses into, through and out of children's lives. As I argue, childhood studies scholars have been fairly slow to deal specifically with (actually pretty vital) questions of energy, in favour of systematic analyses of children's environmental interactions, learning and behaviour. Moreover, different forms of energy – or what I term 'energetic phenomena' – pose a problem for many kinds of (new) materialist thinking (including to an extent OOO). They are not haeccetic, bounded, 'objects' like toys, or animals, or even the macroplastics visible in oceanic vortices, even if one accepts that the boundaries of these material things are porous and blurry. They may not take material forms at all; yet energies, and energetic phenomena, may have material effects and affects; they may interface with (parts of) humans or other machines and media (as Bryant, 2014, would have it); and, building on the previous chapter, energetic phenomena are and have been vital to the infragenerational constitution and the *differentiation* of lives, societies and economies, especially in the Minority Global North. Thus, Chapter 6 offers an account not only of the multiple energetic phenomena that flow through young people's lives – from gravitational to embodied to solar energies – but of how they operate intersectionally (or extra-sectionally) in the contexts of differently positioned young people in Birmingham and Brazil.

Building on previous chapters, I suggest three key ways in which children and childhoods might be analysed with, in and through energetic phenomena: juxtaposition, embodiment and speculative fabulation. Offering a gentle and hopefully generous critique of the climate change movements gathering pace at the time of writing this book, central to all of this is a style of *narrating*, after childhood, that makes space for speculation, circuitousness, listing, playfulness and (anticipating Chapter 9) *silliness*. These, I argue, might be tools for both bringing children and young people (back) into focus and foregrounding opportunities within the seemingly impenetrable workings of global capitalism for those children positioned as 'other' to have a voice or have agency.

Continuing the threads developed in the second half of the book, Chapter 8 offers perhaps the most speculative and 'radical' attempt to think and do, after childhood. I hesitate to use the term 'radical' without scare quotes given what I think a commitment to thinking 'after' entails (see above). However, in exemplifying what interdisciplinary scholarship for childhood studies might look like, it does at least extend far beyond even the forms of interdisciplinarity imagined by critics like Punch (2019) – drawing, as it does, on collaborations with artists and environmental nanoscientists. The chapter asks, more obviously than at any other point in the book, what it might mean to *start-with* a particular genre of material stuff: plastics. In a sense, as I argue, the choice of *plastics* is both important and irrelevant all-at-once, because plastics simply provide a conceptual, political and stylistic entry point (albeit an important one, and one that was of huge contemporary relevance as I wrote the book). Echoing Chapter 3, it then proceeds through various 'cuts': a workshop with school students in which we created totem poles out of wasted or apparently value-less plastic items; looking beyond plastics, to imagine and (via the alien phenomenologies of Ian Bogost [2012]) tell the stories of elements like aluminium and titanium; and (re)focusing on the voices, agency and mobilities of children through hearing more about their everyday engagements with plastics and the other materials that cross through their lives. Beyond an exemplification of interdisciplinary working, the chapter also outlines a set of ways for thinking about plastics and other material stuff, after childhood. The chapter is oriented around two concepts: *synthesis* and *stickiness*. Each idea ushers a number of further implications for thinking and doing childhood research and highlights interfaces and tensions between *silliness* and *trauma* that I explore explicitly in the final chapter. Critically, as I argue, it is no accident that I end the chapter with the voices and experiences of children. For children and young people are always synthesised-with and stuck-with stuff like plastics, and this happens in particular ways, in particular places. Yet, as I hope to show, in order to grapple with intractable challenges – like children's encounters with plastics, metals and other elements – it is always a case of knowing where and when (not) to focus.

The concluding chapter draws together – as far as is possible – some of the most important strands developed in the book. In the spirit of thinking and doing, after childhood, it does not purport to offer a programme for future scholarship, although there are some loose suggestions. Rather, I hope that the book, taken as a whole, offers a series of tools, terms and ways of telling (after) childhoods, which may or may not land in different ways with different readers. In particular, however, and drawing on the work on Berlant (2011) and Braidotti (2011), I examine the possible ethical and political tensions between modes of thinking and doing, after childhood, which are intrinsically interesting, fun, playful or *silly*, and those which are *also* equipped to deal with the (marginalisation, violence or *trauma* of the) kinds of intractable challenges with which I began this chapter.

Notes

1 Whilst speculative realist and object-oriented theories encompass a range of writers and philosophical positions, I use the acronym OOO here and throughout the book for short-hand. For an excellent introduction to the nuances of these approaches, see Gratton (2014).
2 To comply with strict ethical and data management principles, the name of the popular online selling platform cannot be given in this book. Where images from the site appear, for the same reasons, neither the archived web address nor the username of the seller appears.

2 Childhood studies, after childhood

(Post-)childhood studies

I write this book as a human geographer – or as a 'children's geographer', because that sub-discipline has gained increasing prominence and respect over the past couple of decades. My geographical training means I have a particular way of viewing and interpreting childhoods. Evoking earlier work in the sub-discipline, this means an acknowledgement that discourses about (and the social construct-edness of) childhood inevitably involve *spatial* discourses about, for instance, what is acceptable for children to do where, in different societies (Holloway and Valentine, 2000). It also means an attentiveness to where children do what they do – where they play, how they move, what they feel about places (Matthews et al., 2000). And it means analysing how discourses about children, what they do and how they relate with adults, extend across different spatial scales (Ansell, 2009) and operate intersectionally with other forms of social difference (Pyer et al., 2010; Konstantoni and Emejulu, 2017).

However, I also write this book as an interdisciplinary scholar and someone proud to call themselves a 'childhood studies scholar'. I have had the privilege to collaborate with academics working with children from a range of disciplines, including anthropology, sociology, education studies, play- and youth-work. I am also regularly inspired by the literatures – the by-now established journals, books and other outlets – that together comprise the burgeoning interdisciplinary field of childhood studies (what is often referred to as the New Social Studies of Child-hood). The concerns of many other scholars working in this field are my own and mirror those of children's geographers: of recognising that children are, as a result of their generational positioning, legal status, bodily size and/or a range of intersecting characteristics, often amongst some of the most marginalised and silenced in society. Indeed, as the #climatestrikes around the world at the time of writing this book attest, social movements of children and young people dem-onstrate that not only do children's experiences vary with context but there is also something that ties them together in terms of their societal, generational and historical positioning. Thus, like many social scientists working with children and young people, I am passionate about children's rights, agency, voice and politics.

Nevertheless, I have also approached my work with children from at least two critical stances. Firstly, from *also* collaborating with a range of academics who would neither call themselves 'childhood scholars' nor 'social scientists', I have sought to push at the boundaries of what might count as 'childhood studies', at least from a social-scientific perspective. As I detail below and throughout the book (especially in Chapter 8), doing research with engineers, ecologists, ergonomists, architects, artists, environmental nanoscientists and planners forces one to reflect on the established doxa of any field of study – and especially the pervasive "wall of silence" that still seems to exist between bioscientific (especially psychological) and social-scientific (sociological) approaches to childhood (Thorne, 2007: 150), education (Youdell, 2017) and other related matters.

Secondly, and in part as a result, I have for many years been inspired – but also sometimes frustrated – by the exciting, provocative, challenging, generative possibilities of a range of forms of what I termed above 'post-'theorising, for doing childhood studies. Some of these advances have directly broached the 'bio-social' divide, in what Ryan (2012) terms a 'new wave' of childhood studies. Drawing on Rose (2001), it is argued not only that children have been subject to surveillance and regulation by political technologies for decades (Wells, 2011) but that novel combinations of data, technology and psychological sciences mean an intensification of the blurring of boundaries between the 'bio' and the 'political' that can have insidious effects for children's lives (Lee and Motzkau, 2011). In turn, Ryan (2012) argues that bio-social entanglements are not entirely new and warns that attempts at 'hybrid' (Prout, 2005) forms of childhood studies may fail to overcome entrenched disciplinary boundaries.

It has to be said that a concern for the biosocial has been somewhat of a microcosm for a much wider body of scholarship within and at the boundaries of childhood studies. As well as notions of biopolitics, that work has been situated within a diverse range of philosophical approaches, including actor-network theory, assemblage, (feminist) new materialisms, post-humanisms, and non-representational theories. Rather than seek to overcome any biosocial divide, that work has – if it can be characterised by any common threads – sought to various degrees to complicate, question and move beyond the established principles of the social studies of childhood. Overwhelmingly, this has meant an attentiveness of some kind – albeit varying in scope and intensity – to non-human actants (animals, technologies, plants, but especially *material things*), in and of children's lives. In one of the earliest expositions of this approach, Prout (2005: 4) called for an (arguably slightly limited) interdisciplinary study of childhood that examined

> the crucial role played by material artefacts in the construction of contemporary childhood. Childhood is to be regarded as a collection of diverse, emergent assemblages constructed from heterogeneous materials. These materials are biological, social, cultural, technological and so on. However, they are not seen as pure materials but are themselves hybrids produced through time.

Thus, in his analyses of interactions of information technologies (albeit very different in scope and intensity from those discussed in Chapter 5 of the present volume), psychopharmaceuticals and genetics, Prout advanced a language of hybridity, entanglement, becoming, assemblage and heterogeneity that has remained influential.

Indeed, writing 15 years later, it is interesting to observe that a good proportion of scholarship on children and non-human actants focuses on more obviously bounded things – such as animals or toys – rather than the more elusive materialities that Prout examined (but see Gallagher, 2019; Horton et al., 2015; Hadfield-Hill and Zara, 2019). In some senses, then, although the examples, contexts and languages I deploy are vastly different, this book represents something of a return to those *kinds* of concern. However, the rise of feminist new materialist and post-humanist theorising in childhood studies has arguably been just as influential – if not more so in the late 2010s – than Prout's work. It is worth noting that such scholarship remains a subset of the wider field of childhood studies, even if it is growing, and that, quite often, conversations between (for instance) new materialists and others in childhood (and youth) studies can sometimes remain rather awkward (see Kelly and Kamp, 2015, for a particularly insightful analysis of this issue).

Nonetheless, scholarship inspired by new materialist and post-humanist theorising has grown fast. Its key premises overlap with Prout's (2005) work. Empirically, there exist probably thousands of studies now that witness the presence, mutual articulation and entanglement of non-human (or more-than-human) actants with/in children's lives. From rabbits (Taylor, 2019) to institutional materialities (Bauer, 2015; Hansen et al., 2017), from playground materials (Maclean et al., 2015) to stones (Rautio, 2013) and from toys (Lenz-Taguchi, 2014) to bricks, sticks and buildings (Merewether, 2019; Mycock, 2019; den Besten et al., 2011; Kraftl, 2010), these studies bear witness to the sheer variety of stuff that is constitutive of (albeit most often Western) childhoods. Methodologically and stylistically, these studies are replete with attempts to 'work-with' childhoods-natures-materialities and to engage languages that attest to those encounters and forms of embodied practices: manifold forms of observation and (auto)ethnography (Horton and Kraftl, 2018) that examine everyday doings, such as running and rolling (Hackett and Rautio, 2019); attempts at 'being-with' or 'walking-with' or 'becoming-with' heterogeneous childhood assemblages (Kullman, 2012; Blaise et al., 2017); the telling of small, complex, unexpected and emergent stories (Taylor, 2019); and nods to the work of artists and the use of 'post-qualitative' approaches that disrupt traditional attempts to elicit children's 'voices' through experiment (Änggård, 2015; Pacini-Ketchabaw and Clark, 2016; Nordstrom, 2018). Philosophically, this work has been inspired by a range of largely feminist theorists concerned with questioning the status of the human (and the researcher) and with thinking through the ways in which co-minglings of human/non-human might challenge entrenched (white, male, Western) ways of intervening in and thinking about the world (e.g. Haraway, 2011, 2016; Bennett, 2010; Braidotti, 2011; Barad, 2007). At their most radical, these approaches do not simply seek to 'add' non-human actants into the equation but, rather, to decentre humans (and children) in analyses of "sticky knots", "contact zones" and encounters that emerge "when heterogeneous species are unwittingly thrown together in colonised worlds" (Taylor et al., 2013: 53–54; also Spyrou, 2017).

Although similar in their philosophical bases, non-representational approaches to childhood – particularly to children's *geographies* – have also extended in different directions from new materialist and post-humanist theorising. In fact, originating at around the same time as Prout's (2005) book was published, the key contribution of non-representational theories is that they are not only concerned with *matter* (or, indeed, with biosocial dualisms). Rather, their starting point is those aspects of the world that occur too fast for, or are too complex for, or are not irreducible to, representation. The 'non-' in non-representational theory is sometimes replaced (again) with the prefixes 'more-than' or 'beyond' – and each is problematic in its own way – but is intended to afford a sense not that the world *should* not be represented (whether by our brain's ability to process it, or our skill at talking/writing/drawing it) but that a focus on representation may mean that much of what happens in the world is missed. Thus, non-representational work in children's geographies focused as much on affect – on shared emotions and atmospheres, on embodiment (inspired by other feminist literatures) and on querying the apparently linear, teleological logics of 'growing up' in favour of messier, non-linear notions of 'going on' (see Horton and Kraftl, 2006a, 2006b, for more details).

Thinking and doing (and critiquing), after childhood studies

It is important to recognise at this juncture that neither non-representational, nor new materialist, nor post-humanist theories have been immune from critique. Some have accused new materialists of over-emphasising the intentionality of both non-humans and 'data' in research processes (Petersen, 2018). Others charge that they omit the social processes that mediate the ways in which we (as humans, and particularly as scholars) think about the world (Rekret, 2016) and risk viewing the world (and humans) as part of ceaseless flows of energetic forces that risk (re)introducing forms of universalising thought that ignore gendered, racialised or other power relations (Leong, 2016). Meanwhile, criticisms of non-representational approaches to children's geographies have suggested that their attentiveness to banality and forms of experimental, auto-ethnographic modes of witnessing the world are introverted and divorced from the political concerns surrounding children (Mitchell and Elwood, 2012). Some of these critiques have been rebutted (e.g. Kraftl, 2018); others do not necessarily account for the diversity and especially the politics of some new materialist styles of thought, particularly as they have been deployed in efforts to decolonise early childhood pedagogies (e.g. Pacini-Ketchabaw and Clark, 2016; Nxumalo and Cedillo, 2017).

As a result, as with OOO, I am heavily inspired by but not wedded to these approaches, drawing upon a much wider range of scholars in this book. Many of these are introduced at appropriate points in the text. However, with critiques of new materialisms and post-humanisms (in particular) in mind, I want to introduce three bodies of work that have a substantial – albeit not always obviously direct – impact on my arguments. These three strands of scholarship are speculative realist and OOO; queer and critical race writing on materiality; and, theories of generation. I begin with speculative realism and OOO since the overlap with forms of new materialist, 'post-'childhood thinking is perhaps clearest, even if the way in which materialities are figured appears more extreme.

Speculative realism and OOO

As I outlined in Chapter 1, one of the pillars supporting my preference for the term 'after' is the way in which that term has been used by speculative-realist and object-oriented theorists of *matter*. A key commitment in OOO is the idea that major (Western) philosophical traditions remain wedded to the principle of correlationism: that is, that the world only exists for humans, and as we perceive and interpret it. Thus, even the 'post-'theories discussed above place humans at the centre of the world because the world only exists such as it correlates with humans' ability to know it. There is, to paraphrase a range of post-structuralists (but especially Derrida), no 'outside' to human experience. Whilst it is very difficult to argue against this belief – to become an *anti-correlationist*, as many OOO theorists purport – a couple of quick thought experiments may convince that the correlationist pledge is both morally and philosophically problematic (but, as I argue in relation to critical race theories, that is not to say that we should move beyond, only *after* that commitment). For instance, we know that other animals exist: we perceive them, we measure them, we write about them. Moreover, thanks to modern scientific techniques like sonar detectors, we even now know something of how they perceive the world. Yet why do we not take one philosophical (and moral) step further – or *after* that – to ask what it is *like* to be a bat, or a bird, or a slug (Bogost, 2012; see Chapter 8)? And why do we rarely ask what happens when animals relate with one another when humans are absent, or not there to see it – are we really that arrogant and self-centred to assume that *nothing* happens without our say-so?

   Clearly, the answers to these questions are complex and difficult and detain many of the key thinkers on OOO and related post-phenomenologies (e.g. Harman, 2010a; Bennett, 2010; Bogost, 2012; Morton, 2013; Bryant, 2014; Ash and Simpson, 2016). One answer is that, whether intentionally or unintentionally, we are not *prepared* to 'decentre' (or whatever the term should be) ourselves from the world. We have a number of very good reasons for our own self-interest; however, as especially deep green environmental campaigners have been arguing for decades, and as some OOO theorists repeat, humans are *not* the centre of the world, and we share it with many others. Another answer is that, despite our efforts at world mastery, we do not have and will probably never have the tools to account for how non-human others 'experience' the world (even 'experience' being a problematic term because it is bound up in human forms of proprioception). On this, OOO theorists, and especially Graham Harman (2010a, 2010b), have a remarkably simple and striking message: there are many facets of the world that will likely forever remain 'withdrawn' from human experience. Indeed, when it comes to the interaction or (as per Chapter 5) 'interface' between any two objects, those two objects will never fully 'know' one another.

As Harman (2011a: 136) puts it:

[a] raindrop does not make contact with the full reality of the mountain, and neither does a snowflake, a gust of wind, or a helicopter crashing into its face. All of these objects encounter the mountain-object only in some translated, distorted, over-simplified form.

Harman (2011b: 150) does not refute that objects can relate to a degree: we humans and other objects are able to "bathe in them [other object still] at every moment". Harman's language is, however, striking, because it differs from and moves beyond the language of 'relations' that is so important to Deleuze and Guattari (1988) and Latour (2005) and most brands of post-humanism and new materialisms, as they are deployed in childhood studies (see below). Unlike Latour, Harman and others argue that objects must also exist – and be understood to exist – *free* of their relations (Gratton, 2014). The mountain exists relationally: it becomes a "local manifestation" (Bryant, 2014: 4) of mountain-ness in any interaction with an-other and becomes known as a 'mountain' through acts of human naming. None of this is in doubt. But, as Harman argues, the mountain does not *just* exist for *any* or all of those or any other possible relations, and especially not just for humans. Perhaps, the mountain is a good example of this because of its sheer size: just as it is impossible for a human to see the mountain in its entirety, all-at-once, it is impossible to know all of its insides without excavating it to such an extent that it becomes something other than a mountain (a pile of rubble). Thus, contra Latour, the world does not just exist (and nor do objects) through relations, and certainly not only for humans. The question remains, then, of precisely how (and, perhaps, why) one might engage in empirical research *after* (cor)relations and after the human. Or, to put it another way, it is to ask what human 'subjects' like human geography (my own) might look like after all this: "(u)ltimately, how do we account for what remains of the intersubjective when any such subject entering into a relation has already been decentred amid the givenness of the world and so cannot form the foundation or origin of that relation to be built upon? (Ash and Simpson, 2016: 57)" These questions are particularly challenging for any materially inspired analysis wherein a named subject position remains in focus – like childhood.

A second thought experiment once again attends to questions of time and a querying/queering of the teleologies associated with particularly Western conceptions of human existence, generation and childhood. This is most evident in Meillassoux's (2010) writings on earthly conditions *After Finitude*. His key argument for a speculative realism that is anti-correlationist is an ancestral one. That is, if we can only trust our own modes of measuring and knowing the earth, then we must also trust those that prove irrevocably the presence of life on earth before humans (one that, incidentally, corresponds with several religious beliefs beyond or in parallel with theories of evolution). Meillassoux cites the discovery of fossils that pre-date humans' appearance on earth as evidence that the world existed without and anterior to humans and is likely – as Chapters 7, 8 and 9 exemplify and provoke – to continue without us, even if we potentially accelerate our own demise.

Thus, earth, life and all the stuff that comprises it has existed without humans and without our knowing them – even if we *now* know them (and this) (also Nolt, 2004). In other words "the use of radioactive isotopes to carbon date arche-fossils from prior to human consciousness is not in need of any supplement . . . it stands for a temporal proof of a non-correlated reality" (Gratton, 2014: 44).

As Gratton points out, this logic is not anti-human(ist) but an understanding that the world can do without humans. Perhaps most importantly, whether or not one ultimately wishes to take a fully anti-correlationist perspective, Meillassoux offers the clearest explication of the provocative and challenging work that the term 'after' can do: because it attends to earthly conditions *after human finitude* it can scroll both backwards and forwards in time – to that which was anterior to (Modern, Western) humans and that which will inevitably follow, destroy and/or supersede it (see Chapter 9). Although there are obvious limitations for applying this logic to childhood studies, in Chapter 6, I nevertheless extend this argument to pose questions about the temporalities, spatialities and materialities of *child-hoods* that we can only know through material remnants somewhat akin to Meillassoux's fossils. With this tension in mind, I emphasise again that my use of the term 'after' draws on but extends in many other ways than that of OOO theorists. As I argued in Chapter 1, it helps afford a sense in which post-thinking in childhood studies has both gone too far and not far enough – conceptually and methodologically. A consideration of queer and critical race theories, and their critiques of attempts to go 'beyond' (and 'after', as posited by OOO thinkers), provides an important frame for the development of this argument throughout the book.

Thinking sexuality and race, materially

In my view some of the most important and generative critiques of new materialisms – and hence some of the most productive theorisations *of* materialities and their co-implication with questions of *human* difference – have been developed by queer and, especially, critical race theorists. Interestingly, the overlap between new materialisms, post-humanism and, especially queer theory means that there is no clear 'dividing line' – because, for instance, work by Haraway (2011), Chen (2011) and others seems to transcend these neat conceptual compartments, not least as they too deploy certain forms of feminist thought. Thus, in what follows, I highlight as much a series of starting points as critiques for thinking and doing, after childhood, whilst noting that the vast majority of these texts do not have *childhood* as their major object of study.

As I do at various points in this book, in citing and seeking to take inspiration from this work I acknowledge my position as a privileged, white, heterosexual male scholar. Similarly, I recognise that simply stating this is not sufficient, although I hope that my engagements below and throughout the book appear as genuine and meaningful as they were intended. This is because, alongside non-representational, new materialist and post-humanist theorising, I have been inspired for many years by theorists such as bell hooks (2003), Mel Chen (2011) and geographically oriented scholars of youth and race, such as Anoop Nayak

(2010), Sarada Balagopalan (2019), James Esson (2015), and Kristina Konstantoni and Akwugo Emejulu (2017). This is also because, at various points in the book, questions of race and other intersecting forms of 'difference' (although rarely sexuality) come to the fore in the empirical examples I discuss (most obviously in Chapter 7). It is also because, whether writing on childhood directly or not, they urge a generative form of attentiveness to how race, gender and, in some cases, sexuality intersect with other forms of human and (in Chen's and Nayak's cases) non-human difference and with the questions about materiality that I raise in this book.

These forms of attentiveness are vital if we take seriously (and I do) critical race scholars' critiques of new materialisms. The idea that new materialisms extend 'beyond' the human risks (re)introducing a form of European transcendentalism that ignores particularly black (and, more generally, Other) viewpoints and praxes (Jackson, 2015). Thus, whilst promising, perhaps, to go 'beyond' normative notions of personhood, and despite critical scholarship on objecthood and race (see below), "the resounding silence in the posthumanist, object-oriented and new materialist literatures with respect to race is remarkable" (Jackson, 2015: 216). The effects of such absences and silences are profound. As Leong (2016) argues in an equally powerful critique of new materialist theorising – and especially that centred around notions of the Anthropocene – it is vital to call out that *white* imperialist slave-owners were often the drivers of the colonising processes that continue to drive contemporary social and environmental problems, and to note that issues around genetic manipulation, climate change and toxicity – and their effects in terms of racial justice – were being identified long before either new materialisms or ideas of the Anthropocene were put forward. Moreover, Anthropocene discourses are not necessarily divorced from racisms, however unintentional: as Leong (2016) shows, in the longer historical purview, attempts to delineate *geological* epochs have been entangled with efforts to delineate *human* races. Thus, the disavowal of race is actually a structural precondition of the new materialisms.

These are stinging critiques, and I am neither equipped to fully follow through on their implications nor prepared to offer a 'defense' of new materialisms (that would miss the point). Rather, I would make three more focused observations, which weave through the analysis that follows. First, even if they do not fully broach the temporal and spatial scale of the racial injustices outlined by Jackson, Leong and others, there are (albeit tellingly few) examples of new-materialist-inspired studies of childhood and youth that grapple with questions of race and forms of intersecting social difference beyond age (and to an extent gender). Here, Nayak's (2010) evocative and troubling account of how 'ordinary' features of peripheral urban estates are enrolled into acts of white racism is exemplary, as is the work of some Common Worlds scholars in unpicking questions of race, indigeneity and environmental change (Taylor, 2013; Pacini-Ketchabaw and Taylor, 2015; Nxumalo, 2018). Indeed, the feminist underpinnings of new materialisms mean that – at times, but arguably far from commonly enough – some of its key proponents engage with questions of race as well as gender (e.g. Braidotti, 2011).

Second, upon reading Jackson's and Leong's critiques of new materialisms, *my* immediate response – as a white, male childhood studies scholar with a (sometimes albeit hesitant) commitment to non-representational and new materialist approaches – is to ask the same questions about *childhood* as a form of social difference. Indeed, in a sense, although I have not articulated it this way much thus far, the analyses in this book are leading me to this point: whereupon what I term later on the 'pull focus' or arts of *not* noticing might offer a powerful way to expose some of the ways in which childhoods and children are silently enrolled into environmental, material, social and/or digital injustices in ways that new materialisms do not fully consider. I think here, for instance, of Punch's (2002, 2019) long-standing concern that childhood studies lacks a unique conceptual or political voice – where, for instance, one might critically reread the new materialist texts cited in the previous section to ask (however scholarly they are) what in each case is particular to *childhood* about the social materialities they witness. Thus, as I argue below, I agree with Punch (2019) that a focus upon generational orderings and (in)justices might be part of a response and posit my theorisation of infra-generations as just one possible way to work from such a focus whilst retaining *some* attention to both materials and childhoods. I also think that Cindi Katz's (2018) wonderful but provocative characterisation of impoverished children in the Majority Global South as the 'waste of the world' comes perhaps the closest to calling out how *children* (and, notably, often *black* children) are equally or perhaps doubly ignored in materially inflected environmental debates (compare Leong, 2016). Thus, although this is *my* specific reading – and although I am to an extent taking these literatures elsewhere – I want this book to demonstrate how critical race theorists' critiques of new materialisms might *also* – without, I hope, diminishing their message – offer grounds for thinking and doing, after *childhood*.

Third, and a key step along the way, is a consideration of how critical race (and queer) scholars have themselves theorised materiality, often without reference to new materialist or post-humanist attempts to go 'beyond', or to the (anti-)correlationist twists and turns of OOO, but in ways that could work alongside and in productive tension with them. Having already cited several of these theorists along the way earlier in this chapter, I want to focus here on three key terms – *stickiness*, *toxicity* and *litany* – that are particularly important to starting-with plastics in Chapter 8, and to my wider view of the interface between silliness and trauma in Chapter 9, but that also emerge in various other parts of the book.

On *stickiness*, I find Vanessa Agard-Jones' (2012) essay on the ways in which Caribbean sands 'remember' particularly striking, not least in its resonance with Neimanis' (2017) treatise on how water remembers. Agard-Jones (2012: 352) argues that sand is "a repository of feeling and experience" in that – a little like plastic – it can stick to the body and irritate it, but it can also rub off, meaning that sands – for instance, in the form of a beach – may be devoid of *physical* traces of their interface with humans. Like the plastics that simply pass through the human body, sands may be indifferent to the ways in which human bodies 'bathe' in them (to cite Harman [2011b: 150]). Indeed, the resonance with OOO, and Harman's work in particular, is striking, even though I wish to hold it in productive tension here and acknowledge the very different intent of Agard-Jones' piece – which, after all, is to undertake a "queer of colour" theorisation of queer presences on two of Martinique's beaches (Agard-Jones, 2012: 326). I note, however, that Agard-Jones references sand's elemental qualities, which, as sand runs through fingers or becomes stuck to dancing bodies on the beach, offer a glimpse – and only that – of past mountains or coral reefs long ago denuded and, literally, *withdrawn* through the processes of deep geological time. Or, in a similar vein, she references the black volcanic sands from an eruption on the island in 1902 that calls forth associations with queer practices that were taboo then – "as a nod towards a *mountain* of things that could not be referenced directly" (Agard-Jones, 2012: 330; my emphasis is intended to highlight Agard-Jones' reference to the volcano itself, but also to Harman's [2011a] reference to how neither raindrops, snowflakes, nor helicopters crashing into a mountain can exhaust its being). Finally, referencing Jack Halberstam (2005), Agard-Jones (2012: 328) posits the "scattered archive" and a non-normative "scavenging" as a method for witnessing what sands remember. Certainly, this approach sticks with me: it informs most directly the assembly of quite diverse narratives and examples in several chapters.

On *toxicity*, Agard-Jones' (2013) work, alongside Mel Chen's (2011), is helpful for accounting for the presence, circulation and impacts of pollutants and their complicity in what Balagopalan et al. (2019) theorise as the entangled, multi-scalar, diverse forms of unfreedom writ by colonial legacies. It might also, perhaps, account for the generational injustices written into young bodies by air pollution in Denmark, the United States and elsewhere, the kinds of 'resource-power' outlined in the next chapter or the forms of marginalisation and illness experienced by previous generations of children (Chapter 6). In Agard-Jones' case, and in a version of the 'pull focus', she zooms out to consider the implication of black bodies in systems of global power whilst zooming in to consider the micro: "not only [so] that we do research *in* small places but also [so] that we look *from* them – a shift from looking *at* a small unit to looking *from* its place in the world" (Agard-Jones, 2013: 184). Not only is this reference to the 'unit' redolent of another OOO theorist whose work I discuss in Chapter 4 – Levi Bryant – but it offers a racially inflected take on Taylor's (2019) injunction to tell small stories about childhood in the Anthropocene – both arguments I pick up on later. However, it is on the effects of a chemical – chlordécone – that her analysis of toxicity is so powerful – not least because of oblique references to childhood (and fertility) and to endocrine disruptors (with similar effects to, and, perhaps in Bogost's words, metaphorical for, plastics). Entangled within France's colonisation of the island of Martinique, and the banana plantations that flourished under colonial rule, is the use of chlordécone to kill the banana borer beetle. Although banned in France, the United States and various other Minority Global North countries in 1976, it was not banned in the French colonies until 1990. Now found abundantly in the island's soils and water sources, the chemical is a recognised carcinogen and endocrine disruptor and, like certain plastic(isers), is associated with male infertility, birth defects in children and various cancers and neurological conditions (also Davis, 2015). In a couple of striking passages, Agard-Jones (2013) highlights that the circulation of toxins is never racially (or, for that matter, generationally or sexually) neutral and – quite literally – might raise the spectre of life *after* childhood and after the human (again, see Davis, 2015):

> chlordécone has been the source of an emergent gender and sexual politics on the island, where local suspicions about the contamination's relationship to male effeminacy include new convictions about what some are calling a "genocide by sterilization" and a new hospital-based initiative to document intersex births. . . . "What chlordécone has done and continues to do is to call attention to the intimate ways Martinican bodies are connected to commodity chains, to uneven relations of colonial/postcolonial power, and thus to world systems. Taking a chlordécone molecule as a unit of analysis recalibrates the scale of ethnographic practice, bringing us not only to the body, but also to the chemicals circulating within and beyond it."
>
> (Agard-Jones, 2013: 190 & 192)

I will return to how this striking passage reverberates with elements found in soil and water in Birmingham in the Plastic Childhoods project in Chapter 8, and draw out its perhaps more extreme implications in Chapter 9, but also note that Chen's (2011) work on toxic animacies offers a take on toxicity, race, childhood and the material that inflects on my analyses of toy biographies and 'what lies beneath', in Chapter 6. For Chen (2011: 265), the idea of animacy "is built on the recognition that abstract concepts, inanimate objects, and things in between can be queered and racialized without human bodies present, quite beyond questions". Toxins may form part of these discourses. Chen cites the case of Thomas the Tank Engine toys, manufactured in China, but bought and played-with by (largely) middle-class, white American boys. When it was discovered that some of the paint in the toys might contain lead and that that lead *could*, potentially, be ingested by children, there was a furore that connected variously racialised and nationalised discourses: of threatened masculinity, wherein trains are associated with both gendered play and the march of 'civilisation' across the North American continent; of the purity of (white) American childhoods rendered toxic and therefore (racially) impure by the import of Chinese goods; of the silence (compare Jackson, 2015) of the Chinese (child) labourers painting the trains, and likely exposed in far greater doses to the lead, but made just visible enough to be blamed; of the similar silencing and effacing of Chinese labourers working on America's railways during the relentless push westward as the United States established itself as a nation. Thus, toxic animacies refer not only to specific subjects (like white American boys) or objects (toy trains or lead) but to a "toxic sensorium" that crosses borders between countries, that transgresses life/death and that is ruled by a logic of interspersal, symbiosis and *"synthesis"* (Chen, 2011: 272, emphasis mine; see also the example of Gallagher's [2019] geology of media, in Chapter 4). Thus, ushering in a kind of post-phenomenological (Ash and Simpson, 2016) and decentred analysis very much akin to the kind I support, Chen (2011: 281) concludes: "animacy is a category mediated not by whether you are a couch, a piece of lead, a human child, or an animal but by how you interpret the thing of concern and how dynamic you wish it to be". This conception of animacy – and toxicity – throws the scratched, denuded toys that appear in Chapter 6 into a new light, highlighting something of the intersecting and multi-scalar power relations that constitute any (childhood) object.

Finally, on *litany*, and developing Chen's work on toxicity, I turn briefly to Huang's (2017) assessment of the Pacific Garbage Patch and Chinese 'crap'. In this book, I explore how listing and litany (also after Latour [2005] and Bogost [2012]) might offer ways for charting interfaces between and beyond children and material stuff. Importantly, Huang offers a qualification and questioning of how ANT and OOO theorists use listing and litany, through

> what I designate a "plastic litany", or a list of oceanic garbage. The plastic litany is a reoccurring trope in discussions and narratives of oceanic waste – it seems to be an unspoken rule that anyone who writes about the Garbage Patch is compelled to catalogue it. *Here are the things that have been forgotten, not remembered.* In each instance, though, the individual objects themselves are ostensibly irrelevant: the connective tissue of the list is the important part. What difference does it make, if it all ends up in the Patch as pulverized particulates?

Throughout the book, I attempt to attend to those non-object-like materialities – like vortices and energy – that have so often been effaced in childhood studies. Here, though, the point is that listing is – connecting with Agard-Jones' (2012) analyses of sand, and with critical race theorists' reflections on new materialisms – as much a political as a methodological or conceptual exercise, because it calls to thought that which has been forgotten, or ignored, in attention to the material constitution of toxic matters. Speaking directly of race, however, Huang (2017: 105) concludes: "[t]he durability of race, which, much like plastic particulates, photodegrades but never biodegrades, demands an exploration of how Asian American racialization materializes and circulates even in the absence of readily identifiable human bodies". Once again, I ask what the implications of such a position – of such an absence – might be for thinking and doing, after childhood.

Thinking (after) generations

Alongside ongoing tensions about the extent to which various forms of new materialist and post-humanist thinking can deal with what matters to (diverse) children, another key concern within childhood studies has been with the 'contribution' of the field to wider debates elsewhere in the academy and beyond. Most recently, these concerns were voiced ably and provocatively by Samantha Punch (2019), in a way that also refracts on questions of *time*, in and of childhood studies. In a sense the former concern – with the 'contribution' of childhood studies – is not one directly broached in this book. Indeed, any claim to address that in any more than an incremental way would be against the spirit of the term 'after', as it is understood in this volume. However, perhaps implicitly (but for me, more importantly), this book does address two key 'solutions' that Punch puts forward in response to the perceived insularity of childhood studies: a need for interdisciplinarity and a need to look (again) at children with/in generational orderings.

On the one hand, I hope that this book exemplifies the kinds of interdisciplinary scholarship Punch advocates. Indeed, in many ways this book extends well beyond the kinds of interdisciplinary collaborations Punch envisages – confined, as she is in her argument, to broadly social-scientific disciplines like psychology, education, social work or history. Rather, the conceptual and empirical discussions in this book are guided by several projects, over several years (outlined below), which have involved perhaps more 'radical' collaborations with scholars well beyond the boundaries of childhood studies and the social sciences: engineers, environmental nanoscientists, architects, urban planners and artists, to name but a few. In addition, the book is engaged with debates and examples from a range of disciplines – most notably archaeology and studies of digital cultures and technologies – and with methods and techniques (such as the use of social media harvesting and biosampling) that have rarely, if ever, been used in childhood studies, let alone wider disciplines like human geography.

On the other hand, Punch's (2019) core argument – with which I also have considerable sympathy – is that childhood studies scholars (and others) need to engage in far more thoroughgoing theorisations and empirical analyses of *generational ordering* and intergenerational relations. In tandem, they could (I think should) do so with a far greater focus on contexts in which the majority of childhoods are lived – the Majority Global South (also van Blerk, 2019). There is plenty of excellent work on childhoods in the Global South by children's geographers and others – Punch's own work being a case in point (also Punch, 2002; Ansell, 2016; van Blerk, 2019; Day, 2016; Jonah and Abebe, 2019; Balagopalan, 2019). The argument here, though, is that 'age' as a structural form of categorisation (or otherwise systemic and pervasive form of categorisation) does not receive the same treatment as categories like 'gender' (Punch, 2019). Therefore, childhood studies' insistent focus on *children* as 'beings' means that analyses of the workings of age and generation over time, and children's place within them, have been few and far between (although see Alanen and Mayall, 2001). This point extends beyond a consideration of intergenerational relations (e.g. Hopkins and Pain, 2007) to the ways in which – particularly starkly in diverse Majority World contexts – age operates to structure forms of dependency, aspiration, exclusion and resource (in) access (Huijsmans, 2016; Ansell et al., 2019). Although by no means redressing the balance, and although Brazil is by no means taken as 'representative', this book offers a meaningful and rigorous response – not only in discussing research based in São Paulo state (e.g. Chapter 3) but in directly comparing that work with research in the Minority Global North (Chapter 7).

Where I perhaps diverge with Punch's (2019) argument is in her treatment of generations. This is, in part, because her argument is a call for *engagement* rather than a detailed exposition of what a theory of generational ordering might look like. But reading between the lines, Punch's argument is limited to questions of enablers and constraints, agency and interdependency. To be sure, as Punch argues, these questions require considerable further work and matter, profoundly, to experiences of age. Indeed, in the spirit of thinking *after* childhood I certainly do not want to do away with these concerns, as will become clear. Yet I worry that – in reflecting both on some of the empirical research discussed in this book and on the kinds of 'intractable' challenges cited at the start of this chapter – a focus on these terms will not suffice. Therefore, with questions of generational ordering in mind, a mode of thinking and doing, after childhood, perhaps makes the most sense, because that mode would enable childhood studies scholars (and others) not to dismiss but to supplement and look elsewhere than notions of agency (etc.). It would afford an expanded sense not only of *space* but of *time* – and one that includes but extends beyond notions of 'generation'. This is not merely about questioning or queering the linearity of growing up, which forms of questioning may or may not have reference to 'generations' (Horton and Kraftl, 2006b; Sothern, 2007). Rather, for instance, it signals a literally expanded sense in which (for instance) hyperobjects 'phase' into or out of human experience over far longer timescales than those outlined even by Punch (Chapter 4); or of the ways in which objects' biographies may be far faster than, or slower than, *human* generations but nevertheless inflect critically upon them (Chapter 6); or in which, if we take the critiques of the Anthropocene seriously, a decentring of *human* generations in order to better understand non-human temporalities is required in order to get over our arrogant human-centredness in the face of warmed-up, plasticised futures (Chapter 8).

Conclusion: thinking and doing childhoods, after the 'post'?

The implications of non-representational, feminist new materialist, post-humanist, queer, critical race, OOO and generational theories are many and varied for childhood studies. Indeed, at the time of writing, it is fair to say that despite a groundswell of work in childhood studies that is broadly *materiality* focused, these implications have been far from fully thought through. Although I take these literatures in many new directions in this book – including and beyond a focus on materialities per se – three particular contributions stand out. Firstly, as I have just indicated, I seek – if not always to the letter – to take inspiration from the materially inflected writings and critiques of race and queer theorists to think more fully about how their work resonates with childhood studies. This emerges most explicitly in Chapters 6 and 7, but also in attempts to grapple with what – again drawing on the same theorists – John Horton and I have termed 'extra-sectionalities' (Horton and Kraftl, 2018). Looking beyond easily definable, bounded objects (with, perhaps, an implicit critique of OOO), we sought to articulate the presence of social materialities that were *smearing*, *swarming* and *percolating*. These materialities (and processes) were entangled with children's experiences of marginalisation and racial tension in a North London Borough and hence, to an extent, a matter of *inter*-sectionality (e.g. Konstantoni and Emejulu, 2017). Yet we argued that a supplementary term – *extra*-sectionality – might not somehow 'improve' the former concept, nor supplant it, but offer a clearer sense of *what else* takes part in complex experiences and expressions of marginalisation. Thus, we examined the key presences of duck poo (smearing), rats (swarming) and what children not unproblematically termed 'racist groundwater' (percolating) in strongly held feelings of exclusion amongst a white, working-class community on the very edge of London (Horton and Kraftl, 2018). I return to and develop this concept at various points in the book.

Secondly, writing as a geographer, the book is undergirded by and makes contributions to the ways in which *space*, scale and spatiality (the coming together of the social and the spatial) are theorised. On the one hand, notwithstanding some early engagements with OOO and what Ash and Simpson (2016) term 'post-phenomenology', this book offers one of the first sustained and critical treatments of the spatial tropes that are posited by OOO theorists (see also McCormack, 2017; Ash et al., 2018). These theories form *part* of a language for extending well beyond the 'local' scale. Herein, I offer a series of terms that extend beyond 'scale' (Ansell, 2009) and some of the conventional spatial languages preferred by children's geographers (like mobilities). This language includes terms such as 'nexus', 'cuts' and 'resource-power' (Chapters 3 and 6), 'circulation', 'interfaces' and 'visibility' (Chapter 5), 'energetic phenomena' (Chapter 7) and 'synthesis' and 'stickiness' (Chapter 8). Yet I am also at pains – through thinking *after*, not *beyond* childhoods – to make space for and give voice to those more conventional terms developed so carefully by children's geographers and others. Indeed, I argue that it is critical to do so in order to assess how childhood studies might 'speak back' not only to OOO but to the range of theoretical approaches that I have introduced thus far. My inclusion in the later chapters of a range of mobilities, experiences, feelings and narratives – co-developed, often *speculatively* with children and young people – is an attempt to exemplify the importance of this latter kind of assessment.

Thirdly, questions of *temporality* are intertwined with those around space and scale. Most obviously, as I argue in Chapter 6, the term 'infra-generations' might work in the service of thinking both generations and temporalities, after childhood. Infra-generations are, in part, but not only, about more-than-human generations: they also imply expanded notions of time and of the processes in which human generational orders are inextricably interwoven. Moreover, rather than a distraction from or critique of generation-thinking (Punch, 2019), new materialist and other approaches that query distinctions between children and (non-)human others might – whether conceived as 'infra-generations' or otherwise – actually work in support of the kinds of critical theorisations of generational orderings that are required. Thus, a key contribution of this book is to develop – via the concept of 'infra-generations' – a conceptual language *both* for and in excess of human generations (and the ordinary experiential timeframes those imply).

3   Nexus thinking and resource-power

Cuts through childhood, cuts through the earth

Guaratinguetá, January 2016

The city of Guaratinguetá is located midway between São Paulo and Rio de Janeiro, in the State of São Paulo, Brazil. It is a small university city, located on the main motorway – the *Via Dutra* – that links two of Brazil's largest and most important cities. Flanked on two sides by mountains, the region's economy is diverse – ranging from high-tech and knowledge industries, to rice and eucalyptus plantations, to mountain tourism.

Just outside the city is an 'eco-park'. The park – of which different images are shown in Figure 3.1 – appears from some angles to constitute a pleasant, landscaped extra-urban space, with an apparently ordinary playground adorned with a red, blue and white colour scheme. However, the eco-park is euphemistically named. It is, in fact, a municipal waste and recycling site, owned and operated by the Guaratinguetá City authorities. The eco-park points to a series of interconnected processes and concerns, which, in turn, serve as a springboard for thinking and doing, after childhood. These processes and concerns *cut* across the site and are woven into practices, lives and matters both near and far.

Starting with the image of the trucks: every day, hundreds of trucks enter the site, carrying all manner of waste from the city – wires, sheet metal, half-empty pizza boxes, unwanted wooden furniture, sheets of plastic and far more besides. The trucks rumble along the eco-park's dusty tracks, crunching over splintered wood, diesel engines roaring, warning sirens bleeping as they reverse and then shed their loads into untidy piles of scrap. Throughout the day, the procession of trucks is seemingly endless.

Elsewhere on the site, the land has already been reclaimed – whether for the playground, visible in Figure 3.1, or for a grassy, tree-lined, landscaped park. Other than the workers, there is no one on the site: the playground is empty, silent, save for the background roar of the trucks. Compared with the dusty, noisy truck action just a couple of hundred metres away, the relative serenity and lush, verdant aesthetic of the playground and park seems somehow false. It feels too green; too clean; yet, somehow, raw.

This feeling is reinforced by a literal 'cut' into the earth at one edge of the park. Behind a dirty glass sheet (shown in Figure 3.1), one may pause for a moment

Figure 3.1 Photo montage of the 'eco-park', Guaratinguetá, Brazil

Source: Copyright: author's images

and scan down strata of first vegetation and root systems, then the soil dumped over the waste to conceal it and then layer upon layer of human waste, built up over decades. This glimpse – which feels almost voyeuristic – is a slice through the geological record of humanity. There are food wrappers, vinyl discs, broken limbs from plastic dolls (on which far more later in this book), scrap metal, old batteries – all gradually piling up and being returned (or turned) (in)to the earth. Our detritus becomes an amorphous mass, squashed and squeezed over the years by the tonnes of earth and waste above. Although some of this rubbish will not decompose for decades, the timescale of this waste site – this eco-park – as a geological entity is massively accelerated, thousands of times faster than 'ordinary' processes of sedimentation. And, zooming out for a moment, the eco-park at Guaratinguetá is just one of many enormous waste sites in Brazil, and around the world, in which plastics and other human remains are laid to rest.

This cut into one of the earth's newest landscapes offers a small window onto one of the rather more spectacular incarnations of what some term the 'Anthropocene'. It is exemplary of the "geosocial strata" layered onto and *as* the earth (Yusoff, 2017: 105), to be read alongside cities, quarries and farmed fields or terraces as emerging landforms (Dixon et al., 2018). Critically, these landforms are not analytically reducible to geological models or processes. They combine human and non-human matters ineluctably and intractably (Ghosh, 2019). The waste does not appear from nowhere – and only some households in the city have the privilege to be able to divest 'waste'. The trucks are driven; the bull-dozers scrape, sculpt and shape the polluted earth into mounds and valleys that will later be planted with grasses and trees.

Redolent of the visceral falsity of the eco-park (and many thousands more like it), the very idea of the Anthropocene, in its more visible, tangible guises, is a conceit of (M)an, and of particular kinds of (white, privileged) men (Jackson, 2015; Taylor, 2019). The Anthropocene narrative attests at once to the geo-power of the men whose actions – through colonisation, the introduction of plantations, industrial revolutions and global trade – are visible at scale throughout the earth's systems. It also attests to how those same men (and their descendants) have attempted to initiate ever-larger, technocentric solutions that might see them engineer and build their way out of this earthly 'crisis' (Lorimer and Dries-sen, 2014; Stengers, 2015). Moreover, this narrative – as seductive as it sounds when it comes to landforms like the eco-park – tends to conceal and gloss deeply entrenched forms of power relations and injustice.

Moving elsewhere in the eco-park, then, there are reminders that the geosocial strata of this site alone are themselves cut into uneven patterns. Below the spoil heaps where the trucks dump their loads, in an open-sided warehouse, dozens of workers toil in the Brazilian summer heat, sorting all manner of reusable materi-als. This work may not be *quite* as exploitative as it first appears: they work for a local recycling cooperative and, therefore, take a (small) cut of the profits as part of their wages. This warehouse is a microcosm of the seemingly intractable tensions and *troubles* of the earthly conditions that some name the Anthropocene (Haraway, 2016). It is a reminder that not all of our waste is (re)turned into the earth but part of the ceaseless circulation of packaging and technologies around the world, which prop up food, water, energy and other commercial industries. It is also a reminder that, although the cooperative is an innovative community business, this is hot, dirty, smelly and potentially dangerous work – work which goes on at minimal cost to the food, water, energy and other commercial industries that it props up and whose 'green' credentials it helps to reinforce. As I argue in the book's conclusion, work like this reminds of the *trauma* that is a corollary of such industriousness – traumas that cut into the earth as they do (particular groups of) humans.

Moving elsewhere again, and holding that thought for just a moment, adjacent to the site are reminders of other material flows that prop up human economies and metabolisms. Walking back to the top of the 'hill', where the trucks are work-ing, there is a spectacular view of the Paraíba do Sul river valley, with the pilgrim-age city of Aparecida and the Litoral Norte mountains beyond (shown in Figure 3.1). A rough track bisects the sloping mix of earth and waste before giving out onto the valley floor. There, in dark vivid green, are small, perfectly manicured rice fields. In turn, those fields form part of a patchwork of agricultural landscapes in what is one of the most productive regions in Brazil, producing coffee, sugar, cattle, fruit, vegetables and more. Yet, standing overlooking the valley with fresh waste underfoot, one is acutely aware of how this bucolic vista is tied inextricably to the eco-park: some of the food waste from these rice fields will inevitably end up here; and, more insidiously, who can *guarantee* that none of the pollutants from the waste will, over time, accrete into the soils of the valley floor and back into human food systems (compare Agard-Jones, 2013)? As Bryant (2014: 49) puts it:

> we might think of garbage as something simply disappears when we put it in a dump. Yet when we understand that spillage from that dump enters the water supply, affecting wildlife, and that we eat that wildlife, we come to understand the way in which nothing is ever really thrown away, but rather eventually enters us through other, indirect means. A body, as it were, is *sheathed* in a world.

A final 'cut' through the eco-park reveals billboards and what, beyond, is a newly constructed building for local schools (not shown in Figure 3.1). These infrastructures demonstrate the intention that this site also has an educational purpose. Strikingly self-aware, cartoon-style billboards inform children about the dangers of waste and the necessity of recycling. The billboard in the foreground urges children to 'Cuide do Planeta!!!' – to 'Take Care of the Planet!!!', accompanied by an image of a sickly looking globe, cut into, weighed down with and deformed by deforestation and industry. The question of *education* will figure later in this chapter. However, the point of these billboards is that – with the gaudy playground equipment – they serve as indicators that although this is a municipal waste site, crossed daily by tonnes of material stuff divested by the city's residents, there are *traces of childhood* here, which cut across the space in perhaps oblique ways.

The Guaratinguetá eco-park is a fascinating place because, on the one hand, it articulates many pressing (intractable) concerns for our planet, all-at-once. On the other hand, like many other waste sites, much of this site is not really a space *of* humans: yes, it is sustained, managed and shaped in part by human labour, by machines, by inputs and outputs controlled partly by humans; and it deals with 'our' waste (or, rather, predominantly, the waste of a more privileged subset of the local populace). But this site and the geo-power relations (Yusoff, 2017) that striate it are also constituted by the material properties and movements of wasting matter, hidden-in-plain-sight and so often ignored: sliding, rotting, rusting, amalgamating, oozing, shearing, shattering, accreting, becoming-earth, with different speeds and slownesses (Deleuze and Guattari, 1988; Horton and Kraftl, 2018). As Bryant puts it, "with the exception of ecotheory, we have a tendency to ignore the dimension of *waste* produced in all operations" (Bryant, 2014: 263, original emphasis).

Similarly, this is not really a place *for* humans, and especially not *for* children. How many of us would choose to visit or live on a municipal waste facility? Surely this kind of site presents all kinds of hazards for children? Why would children choose to play here? Yet children (or at least, certain discourses and material practices surrounding *childhood*) are woven rather obliquely into the complex nexus of flows, discourses and materialities that constitute this site: food, water, energy, waste, community financing, municipal amenity management, environmental education, play and more. Taking the Guaratinguetá eco-park as a point of inspiration, this chapter begins the task of thinking and doing, after childhood. It does so by responding to two questions, which are picked up throughout the book. Firstly, what can the figure of the 'nexus' offer to childhood studies, especially where childhoods are cut across by, and cut across, a diverse range of earthly challenges that appear so enormous and complex as to be intractable? And secondly – speaking partly as a geographer – in posing this kind of question, and in choosing to focus on a site like the eco-park: *where are* the children, precisely?

Nexus thinking: telling intractable stories of 'resource-power'

One way of understanding the more immediate, complex challenges thrown up by a site like the eco-park is through 'nexus thinking'. Nexus thinking has become increasingly popular in global policy-making circles over the past decade but has begun to filter into more critical scholarship in geography and related disciplines (Leck et al., 2015; Schwanen, 2018). The 'nexus' has been used as a catch-all term for explicating complex systems and, in particular, systems-of-systems. It also represents a particular way of viewing apparently 'intractable' or 'knotty' problems: from the problem of 'waste' at the eco-park to the complex, intersecting challenges that opened Chapter 1 or that striate my discussion of energy in Chapter 7. Nexus thinking has arisen in the context of questions about the readiness of particular policy or industrial sectors – such as food, water or energy – to respond to environmental and development challenges. Specifically, proponents of nexus thinking argue that the silo-ing of 'solutions' into sectors is both artificial and ineffective; rather, they seek to identify and articulate complex interdependencies between sectors and the processes they channel. For, quite simply, "a nexus is defined as one or more *connections* linking two or more things" (Leck et al., 2015, emphasis added). Understood thus, it might be possible to better understand how, in any given situation, there might be 'trade-offs' between different sectors. For instance, back at the eco-park, one might ask any number of hypothetical questions, such as: at what point does the price of new cardboard or plastics – including the fuels needed to transport those materials to and from the recycling cooperative and a food packaging factory – make it unprofitable to use recycled packaging for mass-produced foodstuffs in Brazil?

It has been repeatedly argued that the majority of work on the nexus is a-theoretical. Thus, in comparison with other forms of relational thinking – such as assemblage or complexity theories – at the core of nexus 'thinking' is a fairly technocratic concern for *resource* scarcities (Leck et al., 2015). I want to tread a fairly careful line here. For, in its insistent focus on *resources*, nexus thinking runs the risk of being overly reductionist, instrumental and anthropocentric (not to mention, specifically, capitalist-minded) in its characterisation of material processes.

However, there are, for want of better words, important forms of pragmatism here that could – although nowhere near yet fully developed – have considerable political and conceptual purchase. At the heart of nexus thinking is a particular kind of acknowledgement of material processes, infrastructures and flows that can have profound and divisive effects upon human lives and livelihoods as they are cut across by age (particularly) as well as race and gender.

I term these differential effects 'resource-power'. There is an almost naïve – perhaps speculative – realism to nexus thinking that might challenge those who of us subscribe to forms of assemblage and new materialist thought with a rather unsettling question. Simply put: if, as thinkers of relations that exceed the human, we are still concerned with *human* welfare and well-being, then surely matters-as-*resource matter*, profoundly, to the species? If so, then an analysis of the multidimensional, complex, nexus-forming workings of resource-power might be a way to deal with that concern.

Although not framed in the same way as the above question, it is this concern with material processes and infrastructures that (sometimes) constitute 'resources' that have underpinned nexus thinking. Thus, rather more straightforwardly, the water–energy–food nexus (or 'WEF nexus') has increasingly entered the lexicon of sustainable development over the past ten years (Allouche et al., 2015). Notably, much nexus thinking has occurred at the large scale and has been top down, led by policy initiatives such as the Bonn Framework (Hoff, 2011). Similarly, attempts to map resource nexuses have centred around the production of large-scale, multidimensional, agent-based models. In the case of cities, scholarship on so-called urban metabolisms has attempted to model the flows of water, energy, food and more into, through, around and out of cities, often remaining at the whole-city scale (Gustafson et al., 2014).

Nexus thinking has attracted increasing academic interest, not least within disciplinary human geography (Schwanen, 2018). As will become clear, there is something attractive – or perhaps seductive – about the ways in which forms of nexus modelling can account for massive complexity across systems that are themselves massive. Thus, such forms of modelling attempt to visualise the sheer complexity and potency of resource-power as it is exercised at scale. Yet nascent nexus policy-making and scholarship have been critiqued (constructively, it should be noted) on a number of fronts: for merely constituting a new 'buzzword' that serves further to obscure the primacy of Minority North–led, neoliberalising, technocratic sustainable development logics (Cairns and Krzywoszynska, 2016); for (ironically) reducing complex systems-of-systems to the presumed explanatory power of the 'water–energy–food' triad, to the detriment of other sectors and processes, such as land, labour, soils, chemicals or health (Ringler et al., 2013); for effacing already existing and diverse attempts to witness, acknowledge and deal with nexuses challenges in, for instance, traditional agricultural practices around the world (Allouche et al., 2015); and for insisting on 'fast' models of mathematical abstraction that efface critical reflection upon issues such as governance, power and responsibility for (components of) any nexus (Schwanen, 2018).

Given the technocratic, globalising development logics that characterise much nexus thinking, and given the above critiques, it might appear that the 'nexus' is a strange addition to a toolkit for thinking and doing, after childhood. Specifically, it might not be clear how the idea of the 'nexus' might help to explain the various 'cuts' through childhood and through the earth at a site like the Guaratinguetá eco-park. Yet, as Walker (2019) and Kraftl et al. (2019) argue, within each of the critiques and, specifically, from the very development of nexus thinking in policy and resource governance milieu, there lie – perhaps counter-intuitively – multiple, generative possibilities for theorising *childhoods*. Critically, a re-thinking of nexus thinking might enable something else, and something more, than what is presently offered by new materialist, post-humanist theorisations of childhood. It might, as I argue below, offer a way of specifying how *resource-power* acts in, through and across social and geographical differences. A rather abstract example of a nexus – whose contingency and apparent arbitrariness are not coincidental – should suffice to illustrate this point.

Figure 3.2 shows a fairly typical attempt to visualise the water–energy–food nexus. In this case – as in much nexus thinking – water appears centrally, because, despite an emphasis on 'connections', water is often assumed to have some kind of controlling influence over nexus relations (Leck et al., 2015; Kraftl

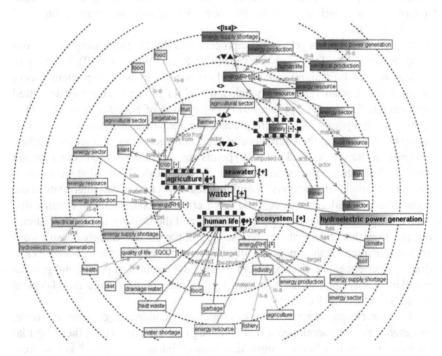

Figure 3.2 A typical visualisation of the water–energy–food nexus

Source: Reproduced with permission from Endo et al. (2015)

et al., 2019). Endo et al.'s (2015) visualisation is based upon their research in the diverse contexts of Japan and the Philippines, and several elements of their diagram are noteworthy. Firstly, conceptually, the aim of a web like this appears to be to *decentre* humans, in an attempt to visualise the complexity of the nexus. Secondly, the web seems to connote some kind of complexity- or actor-network thinking, with non-human and/or more-than-human materialities, infrastructures and processes clearly evident. Notably, some of these nexus constituents exceed the predominant focus in much new materialist-inspired scholarship on childhoods on companion species and 'bounded', haeccetic objects such as toys (Pacini-Ketchabaw and Clark, 2016; Horton and Kraftl, 2018; Gallagher, 2019). Thus – with particular relevance to the eco-park – note the presence of 'waste' (shown middle-bottom of Figure 3.2). Thirdly, and in connection with the first point, 'human life' – below-centre, but off-centre – is not only decentred but reduced to just one (small) component of the nexus, and elsewhere in Figure 3.2 is reframed not in terms of agency (and control) but nexus-embedded processes or resources – such as 'diet' or 'fish resource'. Finally, the web visualises fairly neatly the ways in which nexuses are perceived to work, in practice, through multiple connections and co-constitutions of elements across what become in the process fairly vaguely defined sectors, "in order to maintain their organization and resist entropy" (Bryant, 2014: 101). Thus, similar terms appear across Figure 3.2 rather than in just one place, destabilising their apparently singular use-value and, in turn, their ontological status: for instance, 'fish' appears as 'energy', 'industry' and 'agriculture'.

Nexus thinking is not yet wholly appropriate for thinking and doing, after childhood. Nevertheless, following Schwanen (2018) and Walker (2019), these styles of thinking may offer significant potential – especially in the quite particular ways in which the 'human' (and, by extension, the 'child') is simultaneously decentred and recentred (compare Spyrou, 2017). A key area of potential is *scale* – both spatial and temporal. To be more precise, the analytical and political power of nexus thinking lies at the confluence of scale and *speed*, and to a series of inherent tensions therein. As Walker (2019) points out, a common critique of childhood studies scholarship is its resolute focus on the local scale. Thus, what Katz (2004: xiv) terms "spatial abstractions" – multiply scaled processes that interrelate in the constitution of children's lives – are often effaced (Ansell, 2009). Empirically, such a move is artificial because the world – and the difficult decisions, policies and practices that undergird the exercise of resource-power – simply doesn't work at (just) the micro-scale. But politically, it is even more problematic, because children's experiences are divorced from the broader regimes of power and generational inequality upon which their lives, agency and emergent subjecthood depend (Holloway et al., 2019).

However, childhood studies scholars – including children's geographers – have, in general, not yet developed a requisite language for articulating these 'spatial abstractions'. In this book, I argue that nexus thinking – specified as resource-power – may offer some significant steps towards progressing such a language. Alongside the tools developed in later chapters, nexus thinking requires working through a series of tensions: decentring *and* recentring the human subject; scaling up *and* scaling down; speeding up *and* slowing down. Debates about current approaches to modelling the water–energy–food (WEF) nexus are a case in point. As Schwanen (2018) notes, such approaches are so powerful and attractive, especially to policy-makers, *because* of their speed and ability to articulate scales both small and, especially, very large. Through computational mathematics, they are able to *visualise* (as per Figure 3.2) quickly generated, easy-to-view 'spatial abstractions' – albeit not necessarily of the kind that Katz (2004) had in mind, and of a different kind again than the analyses of Twitter data I present in Chapter 5. They can demonstrate how a confluence of resource processes (say food, water and energy in an ethanol factory) is driven by and in turn drives interactions at other points in space/time (say global energy prices and demand). Thus, (human) 'use' becomes a meeting of scales and is not necessarily 'end use' but simply another node in the multi-scalar spatial tissue woven through a nexus(es) (Walker, 2019).

This is not, however, a simple matter of endorsing modes of mathematical modelling for either resolving resource challenges or for highlighting the analytical deficiencies of micro-centric childhood studies. There are other political and conceptual questions at play. As Schwanen (2018) so powerfully puts it, if nexus thinking is to see through its critical and conceptual promise, nexus scholars now need to dedicate efforts to slowing down these abstractions – but not, necessarily, to supplanting them. Only then, he argues, can questions be raised about the situatedness of the nexus/es: how it/they play out across geographical contexts and historical epochs; and how resource (in)equalities are cut across by the exercise of power through governmental regimes, geopolitical tensions, trade agreements and the status of differently constituted bodies (by age, gender, class, sexuality, ethnicity, religion and/or disability). The question then becomes whether it is both *possible* and *desirable* to scale up whilst slowing down – to articulate abstractions in situ.

Despite an insistence on the local, a small number of childhood studies scholars have – to an extent – sought to extend their scales of analysis beyond the micro-scale. Whilst tending still to emphasise observational and/or experimental methods that witness children's embodied encounters with non-humans (and especially animals), isolated examples of new materialist scholarship make 'bigger' connections. For instance, Pacini-Ketchabaw and Clark (2016) use the figure of the 'water table' in the early years classroom to make connections between children's micro-scale engagements with water and questions of colonial force and water management at the scale of the continental watershed in Canada. Elsewhere, Taylor (2013) engages the 'otherwise worlds' of Aboriginal Australian children, connecting questions of indigeneity, (de)coloniality and non-human natures and weaving dreaming stories with children's material interactions with the desert. Simultaneously, however, these kinds of studies potentially sit in tension with the speed, scale, technicity and instrumentality of most conventional nexus analyses. Although not concerned with nexus approaches per se, Taylor (2019: unpaginated) argues convincingly that not only must our planetary mindsets be slowed down (as Schwanen suggests) but that, beyond the purview of adults, "minor players" such as children are remaking and recomposing worlds. Thus:

> [w]hile other disciplinary fields rush to solve the problem of the anthropo-cene by scaling up efforts to find smarter, bigger and better human inter-ventions, the field of childhood studies is well positioned to scale down, and to cultivate the 'arts of noticing' the small and seemingly insignificant events taking place on the common grounds of minor players. Researchers who draw attention to the *how* of children's world-making with more-than-human others, can contribute to the collective task of refiguring our place in an anthropogenically-damaged world without recourse to the conceits of the Anthropos.
>
> (Taylor, 2019: unpaginated)

On the one hand, this (re)affirmation of the political and conceptual purchase of 'scaling down' is perhaps a clear signal that approaches like nexus thinking might simply reinforce rather than help grapple with the 'smarter, bigger and better' (and faster) 'conceits of the Anthropos'. On the other hand, I want, rather than resolv-ing nexus thinking and new materialist politics into some kind of unwieldy and compromised meta-narrative, to argue for a form of nexus thinking that is alive to both the possibilities and tensions that inhere therein. This means 'staying with the trouble' (Haraway, 2016) – decentring *and* recentring, speeding up *and* slow-ing down, scaling up *and* scaling down – in three senses. Each enacts cuts into, through and across childhood and the earth, and might enable the formulation of critical, eclectic stories about what I term 'resource-power'.

Firstly, as Braidotti (2011) reminds us, the kinds of common grounds and minor players that Taylor evokes are not necessarily engaged in oppositional politics, which themselves carry a range of conceits. For her, oppositional politics "perpet-uate flat repetitions of dominant values of identities, which [they claim] to have repossessed dialectically" (Braidotti, 2011: 40). Redolent of critical race and queer theorists' perspectives on matter and difference, Braidotti values *dissonance* – a stance that requires engagement with, a queer(y)ing of and a kind of experimenta-tion towards apparently dominant modes of storying the world (also Kraftl, 2014). In a striking chapter, Bennett (2001) similarly enjoins us to become attuned to the rather peculiar enchantments of institutional processes, policies and regula-tions. The point, then, is not to celebrate dominant modes of nexus thinking but to unlock their inherent potentialities, to read them dissonantly and to evoke their tensions. To do so is not only to recognise that they are 'not all bad'; they are, after all, seeking to address some of the very same challenges that detain child-hood scholars. Moreover, it is to recognise that – perhaps whatever we do – nexus approaches are very much *on the policy agenda*, both globally and nationally, articulating clearly with the UN's Sustainable Development Goals. The question then becomes one of *how* to work with modellers and policy-makers: of how to be open to *their* decentring of children (and other humans), their voices and agency, but to find moments to recentre them; of how to scale down whilst remaining open to the necessity of scaling up, when issue like the equitable distribution of resources transcend scales; and of how to remain open to speeding up – to the quantifying, visualising, algo-rhythms of nexus logics.

Secondly, then, as Schwanen (2018) argues, the untapped potential of nexus thinking lies precisely in its heterogeneity, ambiguity and instability. Although currently characterised by asymmetric forms of interdisciplinarity, a key challenge for nexus thinking must be to develop theoretical and analytical frameworks that transcend *disciplinary* and sector boundaries (of academia, governments, NGOs and the private sector). With or without the framing of the Anthropocene, the very idea of 'resources' is one that is always-already inherently contestable, complex, unstable and amorphous – taking in not only water–energy–food but waste, societal customs and stereotypes, geology, finance and trade, education, weather, soil, land rights. Just as Yusoff's (2017) conception of 'geo-power' signals an admixture of geologies and socialities, the 'resource-power' that produces any nexus – the multiple energies, flows, channels and blockages – can only be understood in radically interdisciplinary or transdisciplinary ways.

Once again, we (and here I include myself) in childhood studies must to a greater or lesser extent suspend our disbelief: children cannot (always) be central to these modes of inter- or transdisciplinarity; and, more to the point, nor can their voices, agency or politics always be central to flows or blockages of resource-power, even if children are particularly vulnerable to them. What is required is a pragmatic, 'analytical eclecticism' (Leck et al., 2015: 451) that is alive to theoretical and methodological constructs from apparently discordant disciplinary traditions and which may collate (if continue to hold in tension) complex stories about resource-power. One of the central aims of this book is to introduce, develop and exemplify forms of 'radical' interdisciplinarity for thinking and doing, after childhoods. Although perhaps most evident in Chapter 8, I argue in this chapter that a more analytically eclectic, critically aware and perhaps slower form of nexus thinking could be a crucial step along the way.

Thirdly, and taken together, the first two 'cuts' signal a conceptualisation of 'resource-power' as – unlike concepts such as 'geo-power' – something that enables certain kinds of *focus* (and here I do not necessarily mean more or sharper focus). That is, an acknowledgement of the tensions of centring/decentring, scaling-up/scaling-down, speeding up/slowing down, etc. sees particular concerns moving *in and out of focus*. This is of crucial importance to how *childhoods* cut across, and are cut across by, the resource-powers of nexuses, because *children* are particularly susceptible to the workings of resource-power. As resource-power becomes articulated as 'nexus threats' or trade-offs, children are peculiarly and particularly implicated (after Walker, 2019): in the sense that with the longest yet to live (and note, not just as 'future adults'), they have most to lose or gain from particular constellations of resource nexuses; and in the sense that they are arguably most at risk to nexus threats, especially food scarcity and pollution, given their physiological and psychological vulnerability, and their socio-legal status in virtually all contexts (UNICEF, 2014). Thus, Walker (2019) argues that the nexus becomes a significant political-theoretical frame to articulate how children emerge as embodied subjects in and through relations of power at different spatial scales (also Holloway et al., 2019). Therefore, an attentiveness to children's voices and agency is a part – but only a part – of maintaining a sense of processes that take place in but that resonate far beyond children's homes and local communities; a sense of the role of bodies and metabolisms in response to the power of lack; and a sense of how the nexus *policy* agenda could afford greater attention to inter- (or, as I argue later in the book, *infra-*) generational relations, power relations and inequalities. In other words, any (eclectic) analysis of how childhoods are cut across by, and cut across, the resource-powers that constitute nexuses, such as the two that opened Chapter 1, must be alive to the possibility that children move in and out of focus. The rest of this chapter offers two brief and initial considerations for how this might be achieved, whilst later chapters take this logic much further. It asks, once again: *where are* the children, precisely?

Polycentrism: resource-power and education

One of the most potent ways in which resource-power is exercised is through knowledge. Faced with a 'nexus threat' – at *any* scale – one must make a series of judgements: choices; trade-offs; difficult, even impossible decisions, with system effects that may or may not be predictable. In the agricultural regions surrounding Guaratinguetá, farmers, factory-owners and a range of other actors must make decisions about land use, especially in the context of widespread soil degradation. For instance, at a smallholding on the outskirts of the city, one family has taken the bold step to turn a portion of their land over to a tree plantation whose sole function is to retain water and, therefore, improve soil quality on the farm as a whole. There are risks to this experiment: financial investment in an as-yet untested scheme; suspicion and alienation from other local farmers for doing something 'different'; the loss of land and therefore food harvest (in this case, sugar and milk from cattle). There are also opportunities: soil quality has already improved, as evidenced by the absence of termite mounds; despite the loss of grazing land, overall milk productivity has increased; and the farmer is seen (at least by local authorities and academics) as an exemplar for a series of planned green corridors stretching out from the city into the surrounding countryside, supporting water management and wildlife. The family has had to make a series of difficult decisions about the nexus of food, water, energy, soil and landscape, drawing on (and contrasting) different knowledges – from traditional, local farming practices to the advice of regional 'experts' who have supported the development of this exemplar.

The sources of knowledge about any particular nexus – and its constituent elements – are of course diverse. Yet one of the central ways in which the knowledges driving resource-power are channelled is through education – and specifically through education directed at young people. In Brazil, this is achieved – as in almost any context – through the intergenerational transmission of everyday knowledges, skills and habits. At the farm near Guaratinguetá, the older children were heavily involved in farm labour and in the land management experiment, alongside their attendance at school. As Allouche et al. (2015) remind us, these forms of embodied and habituated knowledges should not be effaced in attempts to model 'the nexus'; neither, by that token, should they be romanticised. Nevertheless, in Brazil, knowledges about food, water, energy and the environment more generally are also taught in more formal educational settings: through environmental education.

Brazil is a fascinating case when it comes to both approaches to and, especially, the delivery of environmental education. Generally, Brazilian educators favour dialogical and critical approaches to environmental learning, influenced by the famous pedagogue Paulo Freire (see also Trajber et al., 2019). However, environmental education in Brazil – both in form and content – is heavily influenced by its status as a 'transversal' subject. In other words, environmental education does not form a 'core' subject taught in schools but should 'cut' across those subjects. Moreover, the delivery of environmental education – and the delivery of education more widely in Brazil – is decentred. More specifically, Brazil has a 'polycentric' education system, wherein there are public schools (administered by local school managers), private, fee-paying schools and a vast range of NGOs, commercial companies and other actors delivering education for children (Trajber and Mochizuki, 2015; Feinstein et al., 2013). Following interviews with 64 key professionals working across the food, water and energy sectors in Brazil – including working for the diverse organisations delivering environmental education – the (Re)Connect the Nexus team attempted to map out this polycentric system by tracing some of the interconnections between actors working in the environmental education space in the Guaratinguetá region (Figure 3.3).

During interviews with key professionals, the team asked each actor to explain how they worked with organisations in the delivery of environmental education. In fact, Figure 3.3 shows only *some* of these interconnections. The team identified three 'key' sets of actors, located at the very top of Figure 3.3; the diagram shows only those connections that emerged through our research with industrial (i.e. private businesses) working in the environmental education sphere. Evidently, these connections are many and varied, cutting across scales, encompassing schools, NGOs, agroecology projects (including, for instance, the smallholding near Guaratinguetá), interpretation centres, universities and more.

The diagram is illustrative of an organisational dynamic characterised by *polycentrism*. Indeed, the degree of overall *organisation* in the Brazilian environmental education sector is debateable. There are national environmental education policies and programmes, both drawn up in the late 1980s, with the intention that environmental education should be transversal, youth-led and 'critical'. But the Brazilian system is characterised by complexity, heterogeneity and decentralisation: it is not necessarily dis-ordered, yet the delivery of environmental education involves multiple actors, working with other actors in different ways, ultimately delivering different kinds of environmental education. Thus, in practice, the education that any young person receives is as diverse as the actors or sets of actors involved in its provision.

There has been considerable recent interest in polycentric approaches to governance, particularly in fields of urban planning and energy management – and,

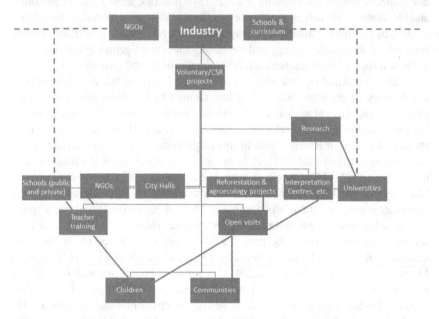

Figure 3.3 Crude mapping of interconnections between different actors and organisations delivering environmental education in the Guaratinguetá region

by extension, notions of 'urban metabolisms' (Taylor et al., 2008; Sovacool, 2014). Polycentrism might be viewed as a particular way of dealing with complex systems of (particularly) resources – as a way of specifying, controlling and managing a nexus. Polycentric approaches to governance (and by extension, the knowledges produced in and for those modes of governance) are "those that mix scales, mechanisms and actors [. . . including those] outside of formal structures at scales ranging from families and firms to regions and intergovernmental organisations" (Sovacool, 2011: 3832–3833). Moreover, polycentric forms of governance "simultaneously [comprise] multiple levels and decision-making bodies, . . . generating and distributing information at multiple scales" (Goldthau, 2014: 138).

In a sense, like nexus thinking, notions of polycentrism constitute a thickly descriptive form of theorising: mapping, modelling and managing connections between actors and across scales. In their different ways, both approaches offer tools for (fairly dispassionately) rendering visible complex connections across previously 'silo-ed' sectors. Yet, like nexus thinking, it would be a mistake to see their greatest value as being the generation of visualisations like those in Figures 3.2 and 3.3. Rather, as I have already argued, the true value of concepts like 'nexus' and 'polycentrism' lies in their exposing *resource-power*. They do not merely describe but urge attentiveness to the forms of governance and decision-making that are emergent from complex systems over which humans do not have total control. In a more 'closed' system, which is arguably not quite as much at the whim of non-human forces – like environmental education – the degree of human control may be greater. Yet the point remains that there is no single, central agent, or organisation, *in control*, and nor is there necessarily any *single* political agenda driving the system. Thus, although power is dispersed and is exercised through millions of diffuse learning interactions, in a subtle but important departure from Foucault-inspired conceptions of micropolitics (e.g. Paddison et al., 2000), these interactions do not necessarily operate in the service of the state. Rather, they represent environmental, moral, commercial and pedagogical agendas as diverse as the agents captured in Figure 3.3.

Under conditions of polycentrism, then, resource-power is diffuse, patchy, heterogeneous and elusive – a little like the materialities of waste that cut through the previous sections of this chapter. However, the forms of resource-power exercised in polycentric systems such as that in Figure 3.3 constitute another 'cut' through the question that forms a refrain for this chapter: *where are* the children, precisely? On the one hand, this question may be answered by focusing on the different forms of environmental education that children receive: for instance, the various messages, aligning with different political and commercial agendas, that might be transmitted to children about how resources can be managed and who 'should' be responsible. These knowledges (and children's receptivity or otherwise to them) would connote millions of instances of resource-power, exercised through millions of micropolitical interactions in classrooms, fieldsites and other learning spaces (compare Kallio and Häkli, 2013). On the other hand, this question may be answered by articulating how children – and notions of childhood – move *in and out of focus* in the organisational logics and pedagogical rhetorics of the differential actors involved in environmental education in the Guaratinguetá region. In the previous section, I argued that notions of 'focus' are important to understanding how childhoods cut across, and are cut across by, the resource-powers of nexuses, because children are particularly susceptible to the workings of resource-power in *general*. With an eye on environmental education, this argument is afforded a little specificity, because we can now ask how children – and how children's learning, well-being and future development – are (or are not) framed by the diverse actors and interconnections of a polycentric system. It is not the intention of this chapter to flesh out these responses. Rather, it is to use the example of polycentrism as one of two brief examples of how children and childhoods might move in and out of focus – whether in analyses of resource-power, specifically, or of modes of thinking and doing after childhood, more generally.

Embodying nexus, embodying resource-power(s): childhoods in and out of focus

Earlier in the chapter, I suggested that (some) childhood studies scholars might need to suspend their disbelief in the consideration of more radically transdisciplinary approaches to thorny problems – like nexus thinking and resource-powers – that nevertheless *matter* to children's lives. I argued, and will argue in the rest of this book, that we purposefully loosen our grip: that we allow for a degree of decentring of both children as subjects (their voices, their agency) and childhood(s) (as the predominant discursive frame of our scholarship). Yet, as I have indicated elsewhere, and will do repeatedly in this book, these forms of decentring would be very far from dismissal; rather, they take us to a place that is *beyond* but not exclusive of notions of voice, agency and politics (Kraftl, 2013a) – even if I now prefer the term *after*.

By way of an illustration of this point, and of a second way in which childhoods might move *in and out of focus*, I turn a final time to the (Re)Connect the Nexus project. As part of our programme of qualitative work with 48 young people, we asked each young person to create a 'visual web' that showed 'their nexus' (see also Kraftl et al., 2019). Although starting with water–energy–food, each web contained a unique combination of resources (and far more besides); each web constituted a unique 'nexus'; and each web represented a unique incarnation of 'resource-power(s)'.

Isabelle (whose visual web appears in Figure 3.4) is an 18-year-old university student studying engineering in Guaratinguetá. Isabelle commutes regularly – often weekly – from her hometown to study in the city. Isabelle's visual web is striking because it not only displays multiple points of connection between water, energy and food in *her* everyday life but clearly articulates a series of *other* concerns. Prominent in Isabelle's web (middle, top), there is a simple image of

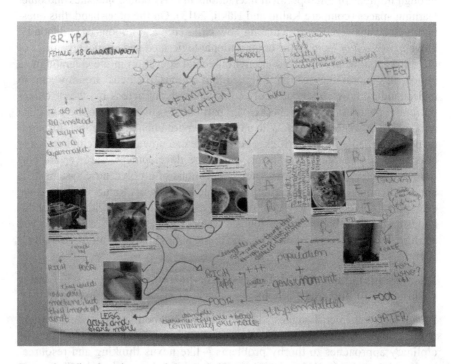

Figure 3.4 Visual web created by Isabelle, an 18-year-old university student from Guaratinguetá

a pedal bike. It transpired that Isabelle had previously walked between the university, her student accommodation and the city centre, where she shopped for food. However, as the simple list (above and to the right of the image of the bicycle) demonstrates, her purchase of a bike cuts across a number of concerns. As a better-off, female university student, she had previously felt unsafe walking from campus to the city centre to go shopping but now felt safer on her bike. The bike enabled her to purchase a greater number of goods in one trip, meaning she could shop at the supermarket – thus saving money and reducing pollution (she had had to supplement walking trips with use of a taxi). And, the bike generally made life easier, because she could carry heavier loads (including her books) with greater ease. Thus, the bike constitutes a signifier, a hub or, in Callon's (1986) terms, an 'obligatory passage point' that forces convergence around a certain issue and becomes a necessary element (or, in this case, technology) through which a network can be formed. Or, to go along with a nexus thinking analogy, rather than 'the' (or even 'a') 'water–energy–food nexus' Isabelle posited a 'food–bicycle–university–energy–money–pollution nexus', with the bicycle (rather than water) occupying a central, controlling position. In this case, then, the bicycle might be taken as an example of an everyday, material, embodied instance of 'resource-power' –here, as a question not of governance (although questions could certainly be asked about women's safety in the city) but of the bike as an *enabler*. Of course, it also implies the entanglement of resources in and as other expressions of power that may feel contradictory: contrast, for instance, Isabelle's feelings of discomfort in urban space with the implied presence of the poorer residents of the city for whom recourse to crime may feel like the only option.

Elsewhere in Isabelle's web (to the middle right), in orange writing on a series of yellow 'Post-It' notes, she notes a second and rather different form of resource-power: 'B-A-R-R-I-E-R-S'. Significantly, as a fairly wealthy student, these barriers (money, education) are not 'nexus threats' for her but represent her concern for other young people living in poverty. Isabelle's choice to acknowledge these barriers is a perfect example of what I suggested above is the *potential* of nexus thinking to move beyond 'thick' description and visualisation towards questions of responsibility, equality and justice (also Schwanen, 2018). In this sense, then, the opportunity to *engage in nexus thinking* (as a creative-reflective *doing*) afforded a striking opportunity to raise questions of social justice. As we have argued elsewhere (Kraftl et al., 2019), Isabelle's choice to call out social injustices in relation to resources (especially food and water) was not isolated. In direct contradiction of many key professionals' stereotype of young Brazilians as increasingly individualistic and materialistic, many young people used their few hours of nexus thinking to expose contemporary injustices in the constellation of resource-power(s) in the country.

José has a very different life from Isabelle, although he lives in the same city. He is 11 and comes from an impoverished background. He lives with his mother and siblings, but his father has left the family home and does not provide any support. His older brother is addicted to drugs and therefore his mother is the sole breadwinner, as well as the main carer in the household. José helps around the house but is too young to engage in paid work. In a familiar refrain, José emphasised particular elements of the nexus as being somehow 'central' or fundamental to his life. Because he often had to go hungry, for him – unsurprisingly – food was a central element of *his* nexus. The detail – and José went into meticulous detail with his drawing – highlighted multiple ways in which food mattered (literally and metaphorically) and in which it was connected to other parts of the nexus (Figure 3.5). Although drawing many pictures of 'special food' (eaten during festivities), he talked most about what is essential (*comida necessária* is 'essential food'). José sometimes cooks, because his mother has some health issues and, alongside working, does not always have the energy to undertake domestic work. He also gets occasional small 'tips' from his grandmother, who lives in the same community, which, he says, he uses to buy 'nice' bread as a treat for the family. Money was central to the ability to buy food – evidenced by the prominent and detailed drawing of a 50 Reais note, which features a jaguar. Yet José also chose to represent other resources and sources of support, beyond the household, through which food could be (albeit tenuously) secured: the support of community members, such as their neighbour, who gives them sugar; and a rather different intersection with education than that expressed by Isabelle – where school was, for José, not only an educational space but one in which he received his only (decent) meal of the day.

José's experiences and understanding of resource-power were at once both intricately detailed and thoroughly embodied. Like Isabelle's experiences of using her bike, embodied engagements with (and as) the water–energy–food (and more) nexus underscored differential experiences of resource-power. Arguably, for José, resource-power was more immediately and thoroughly visceral. Uncertainties over bodily energy – over whether he and his family would have enough food to

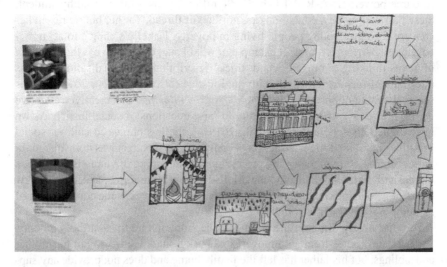

Figure 3.5 Visual web (detail) created by José, an 11-year-old boy from Guaratinguetá

fuel their bodies – were such over-riding concerns that they figured prominently in José's visual web and in the discussion that accompanied it. Food, then, was at the centre of a series of complex nexuses involving school, community, money and temporality (the latter being the rhythms cutting across this nexus, dictating access to *comida necessária* and, on occasion, 'special' food).

Conclusion: arts of (not) noticing

Isabelle's and José's experiences afford an even greater degree of specificity to the theorisations of 'nexus' and 'resource-power' than do the examples of environmental education and the waste site at Guaratinguetá. Their visual webs, and the conversations that accompanied their creation, constituted very particular 'cuts' through childhood and through the often elusive, earthly resources that are vital for sustaining human life. Ultimately, they are indicative of the argument that I want to develop throughout this book that thinking and doing after childhood requires parallel or *juxtaposed* manoeuvres, which can be reductively expressed as 'decentring' *and* 'recentring', although I prefer the idea of 'focus' (see Chapter 7 for greater discussion of the specifics of juxtaposition). Children (and young people) have been a sometimes oblique, sometimes glaringly obvious, sometimes barely missed absent presence throughout this chapter. Children are not and *cannot* always be the centre of our worlds. Yet this chapter has still witnessed a range of processes, materialities, policies and practices in which children are entangled. As I argue in the book's conclusion, despite not apparently being of much or any interest to childhood studies scholars before, these forms of resource-power nevertheless *matter* to children, because they can, especially in the cases of children like José, have potentially *traumatic* implications. It is hard, if not impossible, to witness, understand, let alone intervene into, these matters if we proceed with the conventional approaches that have been developed and mainstreamed in social-scientific studies of childhood over the past few decades. Rather, more 'radical' forms of interdisciplinarity – in which childhood studies might play an inextricable part but may not always take centre stage – are required.

To repeat, however, the visual webs so carefully developed by Isabelle and José serve as a reminder that to think and do, after childhood, does not mean *dispensing* with the approaches that have been mainstreamed by childhood studies scholars over the past few decades – either conceptually or politically. It does not mean ignoring the voices or the social agency of children and young people. Rather, it means carefully weaving, (re)introducing or 'cutting' their voices and agency into analyses of complex, intractable problems – like the water–energy–food nexus and its manifestations as 'resource-power' – at particular times and places. It means bringing children and childhoods in and out of focus, through circumspect and fastidious arts of *not* noticing, or of temporarily suspending our notice, or of (also) directing our attention elsewhere, as much what Anna Tsing calls 'arts of noticing' (Tsing, 2015). For Tsing, these arts are acts not only of careful attentiveness to multispecies encounters, through multiple senses, but also of poesis, of polyphonic narration. This style of thinking and doing is a point of inspiration for acts of juxtaposition, speculation and litany later in the book. But for me – when thinking of and after *childhood* – arts of noticing must also mean looking away: *arts of (not) noticing*. As a narrative and argumentative device, I chose to deliberately 'cut' back to children and young people at the end of this chapter. In my view – and this was my view of things – this was the best way to demonstrate how children and young people are entangled with and productive of nexuses and resource-powers that are complex, multi-scalar, elusive, yet absolutely vital. Clearly, as I argue in the next chapter, other complex, intractable problems might require other forms of bringing children and childhoods into and out of focus.

Engaging in 'nexus thinking' with Isabelle, José and scores of other children and young people, three things stood out about 'their' nexuses. Firstly, and with the other representations of nexuses in this chapter in mind, refocusing on children and young people does more than simply enable us to hear their 'voices' about the water–energy–food nexus. And it enables more than and unpicking of the black box of 'human life' (Figure 3.2). Rather, to recentre children and young people *with those other forms of nexus thinking in mind* is to make a different but complementary 'cut' through nexus ontologies – where stuff like bikes, pollution, money, safety gets added onto or challenges that presumed triadic primacy of 'the' water–energy–food nexus. Secondly, understood and juxtaposed – again – with those other forms of nexus thinking, their webs prompt an important question: although co-produced with children and young people, are their visual webs visualisations of a (highly specified) nexus *around* a young person . . . or do they offer just one 'cut' that exposes the placing of a young person in 'the' nexus (or 'a' nexus) . . . or both? In other words, when children are recentred or brought (back) into focus, to what extent does that centring underscore the supposition that those children's experiences are, also, always-already entangled in relations in which they are not central, to whose functioning they may at times be (co)incidental, yet which matter to their very survival? Thirdly, Isabel and José evoked a range of materialities (bikes, sugar, bread) that – understood at least through the frames of new materialist scholarship – once again evoke the relational constitution of childhoods, querying any sense of this or any childhood as a coherent, individuated subject. Yet there is something about the materialities evinced in this chapter – materialities like 'waste' and 'resources' – that somehow exceeds or cannot be witnessed by new materialist literatures on childhoods, natures and companion species. I return to this point later in the book when I open out just one of the elements of the WEF nexus: energy. There is also something about nexus thinking and resource-power(s) – about the inclusion of governmental logics, policies, professional practices, pedagogies, knowledges, threats and forms of injustice – that *matters*, profoundly, yet with which, again, scholarship on childhood (again, including that of a new materialist bent) has not grappled sufficiently. Thus, and building on the discussion in this chapter, in order to progress ways of thinking and doing, after childhood, the next chapter (re)focuses attention on some of the material stuff of, and *after*, childhoods.

4 Speculative childhoods
Matters beyond materialities

Introduction

The past couple of decades have witnessed a proliferation of modes of thinking about matter. In their different ways, they narrate how material objects are caught up in, produced by and constitutive of human lives. Indeed, many such approaches offer opportunities to decentre humans and/or undermine the idea that humans are individuated, autonomous 'subjects', operating apart and above from the non-human world. Chapter 2 offered a sketch of how different philosophical approaches to matter have been taken up and developed by childhood studies scholars. Arguably, childhood (and, indeed, youth) studies scholars have *always* been interested in materials because childhoods (and, indeed, youth) are so thoroughly suffused with object-worlds: from the meaning-laden homologies of youth subcultures to the sheer attentiveness that children pay to material stuff in their everyday environments and which constantly emerges in (especially geographical) research with children. Nevertheless, since the publication of Prout's (2005) influential *Future of Childhood*, childhood studies scholarship has become increasingly invested in a more limited selection of theories that purport to transcend biosocial dualisms and witness the entanglement of children with non-human, material objects (Lee and Motzkau, 2011; Ryan, 2012). More particularly still, since that time, actor-network theories have been supplanted by new materialist, post-humanist theorisations of *matter*, which – with their underpinnings in feminist theory – have begun to offer alternative ways to theorise the politics of childhood.

Whilst remaining fashionable, and whilst offering much, new materialist philosophies have been critiqued, as I argued in Chapter 2. Nonetheless, in this chapter – and in this book – I want to push new materialist thought further, in the service of thinking and doing, after childhood. As will become clear, this necessitates a double manoeuvre: a healthy dose of scepticism, mindful of the kinds of critiques outlined above; and a willingness to push new materialisms to(wards) breaking point. Indeed, the aim of this chapter is just that – to take materialist and object-theories to their logical limits in ways that might appear wilfully esoteric, illogical, perhaps *silly*, even, but certainly *speculative*. Yet I do so in order to offer a different way to broach the purported exclusion by new materialists

of phenomena such as thought, identity and intentionality. I do so for three reasons: firstly, to continue develop a suite of ways to theorise the kinds of elusive
materialities, nexus complexities and 'resource-powers' introduced in the previous chapter; secondly, to further cultivate the art of (not) noticing – of bringing childhoods in and out of focus in analyses of complex, thorny, intractable
challenges that nevertheless *matter* to children's lives; thirdly, in order to broach
'matter' in ways that both open out new concerns about material stuff and which
raise questions beyond matter, to which I return in subsequent chapters. To do
so, the chapter begins with a fairly lengthy empirical vignette, which hints at the
ways in which new materialist thought might be pushed to(wards) breaking point.
After this excursion to São Paulo, the chapter turns to a detailed exposition of
object-oriented and speculative-realist philosophies, which appear to have literally *nothing* to do with children (or, indeed, any sociological category). There-
after, I attempt – knowing that failure is quite possible – to bring children and
childhoods back into focus through a wide-ranging story about plastics. Throughout the chapter, I offer a critical (and not necessarily celebratory) discussion of
two concepts that might nonetheless lend specificity to my argument: 'machines'
and 'hyperobjects'.

Luz, São Paulo, May 2018

Figure 4.1 shows a street in Luz, in the city of São Paulo. I have been working in
the city and the wider region for the past five years, in a range of collaborative
projects with academics at two of the region's major universities. As part of the
(Re)Inhabiting the City project, we focused our work on Luz. Luz is a neighbourhood located just outside the city's CBD, just a couple of stops on the Metro.
Although not technically a *favela* (or slum), Luz is a very poor neighbourhood,
which faces a number of intersecting problems. It is very densely populated, with
a population of over 11,000 people (although in the 1970s it had a population of
over 40,000 and has suffered rapid depopulation). There is widespread poverty,
with many properties in a very poor state of repair. Many apparently now-vacant
properties in the rather-precarious-looking tower blocks are squatted; the city
authorities take a rather ambivalent view until there is a problem (for instance,
just before our visit in 2018, a large fire killed dozens of residents of a squatted
apartment block nearby).

Despite the poverty of many of its residents, this is one of the city's largest
commercial districts, with informal and semi-legal shops (some visible in Figure 4.1) selling a range of electronic merchandise, of dubious origin. The sector
is male-dominated, employing twice as many men as women. The hundreds
of business owners with a stake in this part of the city have huge commercial
and political power, and have been resisting the regeneration of Luz, seeing
it as a challenge to their businesses. Luz is also known as '*Cracolândia*' –
'crackland' – and is seen as the crack capital of Brazil. This combination of
factors means that Luz is, both socially and environmentally, a very challenging
place to live. Crime rates are high, levels of school attendance are low, and the

Figure 4.1 A typical street-scene in Luz

area is (despite the city being located in a tropical region) subject to increasing levels of desertification.

I want to tell the story of our visit to Luz. I acknowledge, as I did in Chapter 2 when discussing critical race theories, that this act runs the very serious risk of seemingly mobilising a particular aesthetic or affect to make an apparently aloof conceptual point. To do so might be construed as an act of epistemological violence against the residents of Luz, casting them as destitute and de-humanised for the purposes of white, middle-class, Western scholarship. Yet my intention is the reverse, and anticipates my provocations about the connections and tensions between *silliness* (and related sensibilities) and *trauma* (and related effects/ affects) that are dispersed throughout this book and then drawn together in Chapter 9. It is to demonstrate how an apparently aloof analysis of material stuff *is* intimately entangled with/in particular geographical and social contexts; and it is to try to afford a sense – even if doing so is risky – of why such analyses are far from aloof and, rather, entrained in the kinds of questions of resource-power that I raised in Chapter 3.

We (a group of UK and Brazilian researchers) arrive in Luz by car, having been told that it is too dangerous to arrive by *Metro* at Luz station. We walk through the main commercial area of Luz, feeling immediately out of place – largely, it has to be said, because of the area's reputation and the warnings we have been given. Nonetheless, the pavement is broken, with strange liquids running along the side of the road, foaming as they enter blocked drains. It is hot, and humid, and the air smells of burning, and of fast food, and of rubbish, and tastes of dust.

On the street, very small children sit languidly; a pair of mothers lean outside a shop window, feeding their babies; gaunt men sleep in the middle of the pavement, even though it's the early afternoon; and passersby simply step over or around them, hurrying about their business. As we walk on, the contrast between the poverty on the street and the relative wealth of the electronics companies becomes more evident – there are hundreds, if not thousands, of small stores with bright, colourful electric lights, endless arrays of gaudy plastic mobile phone covers, shelves piled high with all manner of (largely un-boxed) electrical equipment, storerooms piled even higher and more chaotically with old circuit- and motherboards, wires, pipes, cables, batteries, and who knows what else. The sheer *mass* of stuff in such a small space – electronic, metal, plastic – is overwhelming.

Our presence is met in a variety of ways. Some people ignore us. Street sellers press up to us, singing at us, commenting on our skin colour, trying to sell us their wares. The few women on the street watch us pass warily. Many people seem simply bemused that we are here.

Making our way past dusty vacant lots used for car parking, we arrive at the Mungunza Container Theatre (Figure 4.2). It is a theatre constructed out of several shipping containers: an initial, precarious and temporary attempt by a local theatre company (Compania Mungunza de Teatro) to change wider public perceptions of Luz and to provide a safe haven for the local population – especially children and young people, as the barrels and other equipment (not pictured in Figure 4.2) hint.

What is particularly striking is the juxtaposition of the theatre with its surroundings. As we arrive, we are greeted by a group of men who flank the metal gates that are visible in this image. They are resting – some sleeping – under the shade of a tree. The smell of alcohol is overwhelming, and many of the men are clearly crack users. The men are wary, but friendly, and open the gates for us. As we enter, the contrast between the theatre's grounds – with its colourful steel drums, plastic geodesic dome and other equipment for the local children to play – and the scene behind is stark. Across the road is one of the city's many Centro Temporário de Acolhimento (Temporary Reception Centres), set up in 2017 to provide refuge and treatment for vulnerable populations living on the street. In Luz, the Centro also constitutes a main drug rehabilitation station, staffed by nurses, doctors and counsellors, occupying a large white tent, and catering for the large population of crack-addicted street residents.

We turn, walk past the theatre, and on our right see two boys playing football on a dusty pitch, with a rudimentary plastic goal. Behind them, the ground floor covered in graffiti, is one of thousands of the city's apartment blocks (Figure 4.3). Some are so thin, with crumbling walls and paint, that they look like they could topple over at any moment.

Figure 4.2 Mungunza Container Theatre (play)grounds, with a *Centro Temporário de Acolhimento* (temporary reception centre) in the background

We enter the building – unsure as to how sound the structure really is – into a dark, windowless, stuffy ground floor space, littered with all manner of stuff: bottles, clothing, boxes, wires, pots.

Heading further in, it becomes clear that the building was once (perhaps just 50 years ago) a fairly grand apartment block for the newly rich lower-middle classes who were moving to the city centre but has since been vacated and then squatted; today it is, as far as we can see, devoid of life, save the pigeons who fiercely protect their nests as we disturb them.

We arrive at a central stairwell, where plants grow out of the walls, whose railings have been removed for scrap and which barely resembles stairs. The building is crumbling constantly; plasterwork cracking away from walls and ceilings, lumps of brick littering the stairs and skittering downwards as we climb up. The building has taken on almost geological feel; the stairways could be scree slopes on a desert mesa.

With trepidation, we enter several of the apartments, each of which has a crudely painted number above the door. It would appear that the last residents – the squatters – were forced to leave in a hurry, perhaps just two or three years previously. I walk into one apartment, startling a nesting pigeon, which in turn causes me alarm. I back out, barely able to navigate my way over a heap of wooden

Figure 4.3 The exterior of a 'vacant' apartment block, next to the Mungunza Container
       Theatre

boards, steep poles, dirty clothing, electricals and other indeterminable stuff. I enter the adjacent apartment, which is covered with layers of dirt, more wood, more clothing and all kinds of paraphernalia. My eye is caught first by an abandoned plastic doll, lying on its back, arms by its sides, its lifeless eyes staring at the ceiling (Figure 4.4). It sits at the entrance to the apartment; on its own, it almost seems to have been placed there (perhaps it has), to evoke a sense of uncanniness, loss or anger. Or perhaps it has been dragged there by a pigeon, or carried by the torrential rainwater and drying winds that regularly penetrate the building's cracking walls and open windows, or fallen through the hole in the roof to the apartment above. Or, perhaps, it cannot be separated from the other matter in this room – clothes, pillows, a child's pink sandal, a splintered, faeces-covered plank of wood – objects that interact with one another, degrade with different speeds and slownesses, part of the generic sliding, rotting, oozing, dusty mass that this edifice is becoming. In turn, this doll, in this room, in this apartment block, is just one of thousands of dolls, of thousands of rooms, of thousands of apartment blocks in this area of the city, which are – hidden-in-plain sight just a couple of *Metro* stops from the city centre – gradually becoming ruinant.

From my vantage point as a privileged, British academic, this tour around Luz – and this building – and especially this doll, and the children's sandals, and a plastic red toy car I found on a crumbling balcony – they all haunt me, as, I suppose, any ruinous, abandoned site might do. There are many ways in which we might respond to such a tour – many of them, I suspect, both fairly emotive and pretty questionable in ethical terms.

But this tour also brought to life (if that is the right phrase) something of what I have in mind in my intention in this chapter to push new materialist thought on childhood to(wards) breaking point. In fact, four more specific questions stand out. Firstly, in a conceptual context that seeks to acknowledge the role of non-human matter in the constitution of human lives: how to make sense of the pressing, massing, rotting, flowing kinds of materialities that struck us on this tour (and which, perhaps, constitute further incarnations of 'resource-power')? Secondly, what is the place of particular orders of materialities – such as woods, metals or *plastics* – within the ceaseless and apparently senseless circulation of material stuff in the apartment blocks, electronics stores and streets of a place like Luz? As I ask in Chapter 8, starting with plastics, how do different kinds of material stuff circulate, (dis)entangle, *synthesise* with and *stick* to one another? Thirdly, having spent nearly two decades working with children and young people, seeking actively to instil their *presence* – their voice, their agency, their rights – how to make sense of a space like the apartment block where childhoods seem (perhaps paradoxically) to have been evidently dis-placed? In other words, what to make of traces of childhoods, and of individual children – dolls, sandals, toy cars – that are entangled with the amorphous mass of this crumbling building? And, fourthly, how to balance what is a (perhaps immoral or at the very least distasteful) fascination with the materialities of this place with an attentiveness to the very real and very pressing problems facing the children who left those toys behind, as they grow up in a place like Luz?

Figure 4.4 The interior of one of the abandoned apartments, with a plastic doll in the foreground

Pushing new materialisms to(wards) their limit? Childhoods and/as machines, childhoods and/as hyperobjects

It is very difficult to settle upon any singular explanation or theorisation of the matters that constitute Luz. The amalgamated stuff in and of the apartment blocks (it's hard to tell which bits were *left in* the building and which *were* the building as it has crumbled) encourages speculation. Speculation about the stories of the lives lived in the building and the lives of those who evidently left in a hurry. Speculation about the energies and forces that brought together to the specific, apparently random assemblages of stuff we encountered in the apartment. Speculation about how those assemblages might be read together with the apparently intractable problems affecting Luz. Speculation can, of course, be a dangerous thing, promoting introspection, frivolity, a lack of rigour, narrow-mindedness and a blindness to working with others to seek evidence or explanation. Yet, as I argue throughout this book, speculation might also be expansive, enabling experimentation, creativity and/or alternative ways of looking at or inhabiting the world: a telling of small (and/or tall) stories that provoke further thought and action and that may be able to broach the kinds of trauma likely associated with *that* doll.

In Chapter 2, in order to sketch out the multiple ways in which I understand the word 'after' in thinking and doing, 'after childhood', I briefly introduced speculative-realist and object-oriented ontologies (OOO). To my mind, these ways of thinking matter are speculative in at least one specific sense: they require a certain leap of faith beyond correlationist thinking – beyond the assumption that the world (and any object therein) exists solely for humans, as they perceive it. In this section, I discuss two such speculative styles for thinking matter, which I believe might – although not unproblematically – offer *part* of a toolkit for addressing the complexities of life in Luz (and elsewhere). Again, I do not wholeheartedly endorse these styles, nor suggest that they are 'applicable' to any or all matters: indeed, they should be read and juxtaposed alongside the other (speculative) styles of thinking I develop in this book. Nonetheless, I turn first to Levi Bryant's (2014) exposition of 'machines' and then to Timothy Morton's (2013) conception of 'hyperobjects'. My reading of both concepts is partial but is intended to open out further possibilities for thinking and doing, after childhood.

Machines

Levi Bryant's (2014) theorisation of machines has a familiar starting point: the de-privileging of materiality – of material stuff – in historical and social-scientific accounts that have been chiefly concerned with discursive practices and representations. For Bryant, as for many philosophers of the material, the upshot is not only that matter becomes a stain or blind spot (Zizek, 2006). Rather, it is that, if things have power (Bennett, 2010), "an entire domain of power [becomes] invisible, and as a result we [lose] all sorts of opportunities for strategic intervention in producing emancipatory change" (Bryant, 2014: 3). As Morton (2013) also argues, a key upshot of this blindness has been an inability to grapple with the material effects of climate change – and, as I outline in the closing part of this chapter, the materialities of other hazards, such as plastics.

Amongst a by-now considerable range of responses to the problem of matter, Bryant's is to develop what he terms 'onto-cartography': literally, a mapping (cartography) of things (onto). Onto-cartography has (in my reading) two foundational premises, which offer a distinct and particularly helpful resource for witnessing the social-materialities of childhood (Horton and Kraftl, 2018). The first premise is, in part, lexical. Rather than talk of 'matter' or 'things' or (at worst) 'objects', onto-cartography is a method for mapping *machines*. In part, and to simplify grossly, this means that onto-cartography has a more expansive understanding of matter than either the 'objects-with-relations' thinking of ANT, or the 'objects-without-humans' thinking of Harman (2011a), or the 'materials-as-identifiable-objects' thinking of (especially childhood studies' articulations of) new materialisms. Rather, onto-cartography is concerned with material 'stuff' *and* material 'things' – with *both* things that present themselves to humans *as* discernible objects and with the amorphous, swirling material massifications of stuff in the shops, backrooms and apartments in Luz. Crucially, things/stuff do not merely relate but *operate*; onto-cartography

> recognizes the existence of discrete, emergent entities [. . .]. [W]orlds are composed of units or individual entities existing at a variety of different levels of scale, and that are themselves composed of other entities. I call these entities "machines" to emphasize the matter in which entities dynamically *operate* on inputs producing outputs.
>
> (Bryant, 2014: 6)

Because they operate – rather than simply exist within sets of relations – it is important to note that the move to specify particular (constellations of) matter 'machines' is not merely lexical but political. Moreover, a focus on machines helps evade the immediate connotation of 'objects', 'matter' and 'non-humans' – that they are distinct from or opposed to 'subjects', 'non-matter (discursive practices)' or 'humans'. This, then, is a further step away from the strictures surrounding 'biosocial' conceptions of childhood, which I outlined in Chapter 2.

The second premise of onto-cartography is that it is – in principle – slightly less critical of correlationist doxa. Bryant is clear that he does not feel that various brands of (for a childhood-specific example) social constructivism are wrong; merely that they overstate the case for the significance or necessity of human representations or discursive frameworks. Rather, onto-cartography might enable the retention of the political or ethical demands of (say) social constructivism precisely because of its commitment to machines rather than objects. For machines do not merely operate but are *programmed* – whether by humans, by genetics, and/or by the ways in which, at an atomic level, apparently inanimate materials such as metals 'conduct' (behave, usher) themselves through quivering, cracking, tumbling movements (Bennett, 2010: 59). What I have termed 'resource-power' relates to just one register in which those programmes occur, and in which humans are often (but not always) in focus. But programmes may also translate or be translated; deploying the language of software engineering, Miller (2013: 4) argues that concepts may be 'ported' from one context to another. Bryant seeks to do the same: to rework the political and conceptual programmes of Marxism, critical race theories and feminisms with/in the foreign environment of onto-cartographies. And *my* hope is a similar one: that porting – which may imply transformation as much as translation – between childhood studies and onto-cartography might be an effective step along the way to thinking and doing, after childhood.

With these two premises in mind, it is worth considering the properties of machines (and, by extension, of onto-cartography) in more depth. As will become clear, an onto-cartographic conception of machines is to some extent a re-programming of other forms of non-dualistic thinking. However, a machinic disposition does enable some particular possibilities for thinking and doing, after childhood. I confine myself to seven such properties, which are synthesised from Bryant's (2014) work alongside that of a range of other cognate scholars. First, machines may be corporeal and/or incorporeal, material and/or immaterial. Second, machines may operate at scales from, or across, the very small to the very large and may be nested within one another in the constitution of assemblages (Anderson and McFarlane, 2011). Third, not all machines are either mechanical, nor are they designed as such. Thus, works of art can be machines, but, because 'creativity' is not the sole property of humans, machines – including metals, as noted above – can evolve, or emerge, or become other without the interventions of 'subjects' (Grosz, 2011). Hence, "[t]he term 'machine' allow[s] us to escape the anthropocentric associations of the term 'object' by drawing our attention to beings that operate as independent bodies . . . only a small subset of [which] is designed by humans" (Bryant, 2014: 18). Fourth, then, not all machines are rigid: whilst a pencil sharpener is composed of fixed parts, it is part of a 'subspecies' of machines where others can grow and develop – "[c]hildren, otters, organizations, etc., are capable of learning and changing their behaviour and operations as a consequence of what they have learned" (Bryant, 2014: 16). This short extract – with its apparently arbitrary listing – speaks to me: politically, for the way in which, *without prejudice*, very different kinds of machines are placed alongside one another; and conceptually, for the way in which *children* appear as both part of an apparently arbitrary list ('etc.') and important enough to name, nonetheless. Whilst Bryant says very little (if anything) about listing, I work up my interpretation of this idea in more detail in my discussion of plastic (childhoods) at the end of this chapter and in Chapter 8.

Fifth, and as a consequence of the first four properties, machines are neither, of necessity, passive nor defined in terms of their 'use'. Citing Lewis Mumford, Bryant uses the example of steam engines, whose ostensible 'purpose' was to create energy to do *something*. The developing capacities of the steam engine were contingent upon the operation of machines working at a smaller spatial scale – not least the characteristics of metals at different temperatures. In turn, the steam engine ushered a range of unforeseen consequences: the creation of engine sheds for the better care of steam trains; the grouping of engine sheds for the more efficient pooling of labour and resources; and, ultimately, the (re)organisation and development of the industrial city around the warehouses and railway lines (as well as a range of other technologies, such as the clock) that resulted, as Mumford (1934, 1967) showed throughout his work. An important implication of the non-passivity of machines is what Bryant (2014: 20) terms their 'insistence'. That is, whilst machines do not necessarily structure human lives – this is not a call for a renewed environmental or technological determinism,

> a tool or environment [may] habitually structure the body. The ink pen calls
> for certain ways of being grasped. Not only does it likely have an effect on
> the form that muscle and bone morphology take over the course of repeated
> and continuous use, but it also generates various neurological schema or ten-
> dencies to grasp that, in their turn, close off other ways of grasping.
>
> (Bryant, 2014: 20)

One may list innumerable other examples of machines that operate in this way – which is ostensibly a kind of porting or re-programming, to return to some ter-minology used earlier. These could range from the changes to the musculature of dogs, chickens, cows and other animals arising from selective breeding and the intensification of farming for meat to the ways in which a lifetime spent working in a particular job (for extreme examples, sat day-long at a computer, or working in a coal mine) that are expressed in the comportment of a body, its callouses, its stresses and strains and, ultimately, its longevity.

The insistence of machines begs two questions. The first – whose response I will defer to Chapter 6 – is a question about generations. That is: how do we think generations beyond the human (both in terms of 'machines' and in terms of their multiple temporalities) – as *infra-generations*? The second question – which I cannot answer in the general here, but which I return to throughout this book – is: how to *respond* to the insistence of machines? How in the world – as human-machines, composed *of* machines, composed *by* machines and compos-*ing* machines – can we choose which machinic operations to respond to, how do we *interface* with (and as) machines and what should be the appropriate and proportionate response (see Bennett, 2010: 122)? Many thinkers concerned with the Anthropocene, its various alternative designations and, as Bryant urges, its material effects are, rather than shying away from these questions in a form of speculative escapism or introspection, engaging with these questions in expan-sive, enabling and experimental ways. Here, as elsewhere, I think of Donna Har-away, Anna Tsing, Isabelle Stengers and a range of others whose writings inspire other styles and modes of response-ability to the insistence of machines that are creating manifold forms of earthly injustice. I will return to this question through-out the book, but most immediately through a consideration of Morton's (2013) 'hyperobjects' and plastic (childhoods), below.

Sixth, Bryant's conception of machines is – in its commitment to *and* beyond matter – dominated by a particular conception of media. Whilst I discuss this in more depth elsewhere (Chapter 5), the relationship between machines and media is one that is pertinent to ongoing debates about childhood, youth and technology. Like Marshall McLuhan, Bryant (2014: 31) develops what he terms a "post-human media ecology" that moves beyond analyses of popular media to an understanding of "intermediaries". Once again, this requires attunement to the material manifestations and operations of media. Parikka (2013: 527) uses the term "medianatures" to specify what this attunement might look like: a broadened conception of media as a combination of earthly energies that produces effects, affects or intensities. Perhaps the most straightforward way to think media beyond the human is to think of soil as a growing *medium* that combines various properties to enable plants to grow. Alternatively, Neimanis' (2017) striking articulation of the hydro-logics of water attests to water's capacities for communication as it literally carries all sorts of chemical and material flotsam and jetsam, whilst seeping into and becoming a medium for (for instance) resource-power. Thus, as Neimanis demonstrates, water may become a medium for social differentiation, dissolving and hybridising human/non-human 'divides' in the production of "hydro-socialities" (also Clark et al., 2017: 1351). Later in the book, I consider how plastics' multiple capacities for *synthesis* and *stickiness* afford rather different kinds of media interfaces and social differentiations.

Bryant's conception of media hinges around an argument that media are not merely prostheses for sense organs but, rather, operate as media whenever they provide the impetus or the energy (like soil and/or water) for the modification of another machine. Media are not, then, only extensions of humans (as McLuhan saw them) but may not involve humans at all – or else humans may be media for *other* machines, such as parasites. This conception of media accounts for what media *do* rather than what they *are*. Neither I nor you is necessarily a medium – although it is likely that we are playing host to innumerable machines – but our bodies have the capacity to act as such. In other words, "[m]achines . . . are not representational, but rather are productive" (Bryant, 2014: 39).

This way of viewing media may go some way towards accounting for the doll I came across in Luz. Clearly, in this case, to view the doll as a popular cultural object – replete with connotations about gendered play, or its marketing to specific groups of children – would be to (at least partly) miss the point. And the point would (at least partly) be a political one. We can only speculate, but I wonder whether this doll, and dolls like it, has been mediated rather differently from those that make their way into the arms of richer children. Perhaps this doll *was* bought new and given to the owner who left it in the apartment. Perhaps it was passed down from mother to child. Perhaps it was found in another apartment. Perhaps it was not left here by its last owner but swept along, with feathers, and dust, and sandals, and shredded wood, by the wind or the rain or by rats. The media that modified this doll remain uncertain but, simultaneously, this doll acts, as Bryant has it, *as* a productive medium, communicating something of the sociomaterial injustices that led it to be in *that* apartment on *that* day. Critics of these kinds of speculation will point (rightly) to the fact that they are centred upon my own encounter with, and reflections upon, the doll, even though we can probably agree that the doll did not and does not exist *for* me. Yet I know that this account – even though the child or children who we presume to be the owner(s) of the doll is/are absent therefrom – is not yet sufficient to convince that it is *after* childhood. There is more work to do to reach this stage, which, I hope, reaches its culmination in Chapter 8, where I turn to Bogost's conception of 'alien phenomenology' (with some further input from Bryant's work on machines) in order to scaffold my work on plastics and childhoods.

Seventh, and building on the previous point, machines embody and exert *power*. In my reading, this acknowledgement is critical to Bryant's political and ethical project and, indeed, to (m)any contemporary theorisations of matter. Following Deleuze and Guattari's (1988) well-known articulation of the virtual and actual, Bryant (2014: 4) distinguishes between the powers of "virtual proper being" and "local manifestations". Virtual proper being denotes the powers of a machine whether or not they are exercised, because, other than in exceptional circumstances, no one exercise of power exhausts the potential of a machine. Some machines – particularly those that are more complex, such as the lumps of flesh that we call 'animals' (including 'humans'), or large multinational corporations, or forests – hold arguably more diverse or substantive potentialities. Others, perhaps, less so. But, note, this does not necessarily ascribe particular kinds of machine with more importance, particularly since the (albeit more limited) potentialities of a particular machine may, in certain circumstances, override those of all others (like the grain of sand that flies into your eye, forcing you to take an impromptu trip to find water to wash it out, making you late for your meeting, with whatever consequences ensue). Hence, part of the point of the Luz example has been to ascribe some operational power to a doll that otherwise appeared worthless, in a place considered abandoned by humans with very little economic, social or political power.

It is also the case – for Bryant (2014: 42) at least – that the powers of a machine are objective: "[t]he powers that a machine possesses are features of that machine regardless of whether or not anyone knows of them or has observed them". Here, Bryant comes closer to the arguably more radical forms of OOO wherein humans recede from view (see Chapter 1). But, in a way, this must make sense for the world to keep working as it does, even if it means admitting that – as many critics of the idea of the Anthropocene have argued – the world does not exist for humans. Local manifestations are not necessarily of or for anyone and their perception thereof. Iron rusts regardless of whether anyone sees it; and it would be ridiculous to assert that the decaying and gathering of the objects in the apartment in Luz happened for me or only happened once I had perceived and registered it. There is a difference between the processes that happened and how I subsequently represent them, even though there is clearly an overlap and an ineluctability – but that does not mean they are the *same*.

Tying this entire discussion of machines together, onto-cartographies of power, then, would overlap with but also extend analyses of the workings of power in societies – if for no other reason than the truism that many local manifestations of power either do not include humans at all or see humans dis-placed from the story. Developing the arguments of the previous chapter, onto-cartographies would be mappings of the waxing and waning of a machine's power, and of the operations of a machine, and of its insistence upon other machines (and the changes that do or do not ensue), including but beyond perhaps anthropocentric notions of 'resource'. If this seems nebulous or even nihilistic (where should these mappings start, or end, and how to discern what is 'good' or 'bad'?), then critical here is a sense that the virtual powers of an object are not formless or unstructured: there are no unformatted beings, as Harman (2010a) puts it. It is not the case that every time power is exercised, it is shaped out of some meaningless or formless primordial soup. The flows of the world, through which powers are exercised, are channelled or halted – in ways that can be vicious or beautiful, progressive or damaging, as my discussion of *resource-power* demonstrated in the previous chapter – depending on what it is that a machine insists of or imposes upon another machine(s), and therefore the positionality of that machine(s). "The consequence of this is that machines, in performing operations on flows or inputs, will have to contend with the powers characterizing the being of these flows" (Bryant, 2014: 50).

Before moving onto a discussion of hyperobjects – and in order to provide an example that anticipates that concept – I want to focus briefly on a childhood-related example, taken from the provocative and innovative work of Michael Gallagher (forthcoming, 2019). In his paper, Gallagher sets out an agenda for research on children and media that looks beyond the traditional scope of such work: children's use of media (e.g. Goodyear and Armour, 2018). Like Bryant, he borrows from Parrika's (2013) work on "medianatures" and post-human media ecology: media "are not only a technology, a political agenda, or an exclusively human theme. Media are a contraction of forces of the world into specific resonating milieus" (Parikka, 2013: xiv). Specifically, however, Gallagher (2019: unpaginated) is interested in what he terms a "geology of media", focused on the technological and earthly manifestations of media – media's matter. Critically, he argues that "[t]he planetary nature of digital media require analyses that, in line with wider arguments in childhood studies, follow hidden relations across different scales, and address the politics of harmful materialities" (Gallagher, 2019: unpaginated). On the one hand, his emphasis upon the scaling and phasing of digital media – their globality and sometimes hidden relations – resonates with Morton's (2013) conception of hyperobjects, to which I attend below. On the other hand, to trace the geology of media is far from an a-political act (something I will argue about plastics as hyperobjects later in this chapter). In fact, the reverse: Gallagher is concerned with how processes that are understood to be part of the Anthropocene – at the planetary scale – in fact have harmful consequences that require a political response. Again, in Bryant's (2014) terms, these apparently geological processes have an *insistence* that needs both attention and resolution.

Gallagher is specifically interested in intersections of geology, media and *childhood*. He focuses on the cobalt mining industry in the Democratic Republic of Congo (DRC), where children are employed in excavating materials that go into mobile (cell) phones. Indeed, cobalt is an element that is found in many electronic devices, including 'environmentally friendly' cars, because it is a key ingredient in fabricating lithium-ion batteries. For various financial and political reasons, a fairly large proportion of the DRC's mining industry is artisanal and informal in nature, and it is estimated that around a quarter are aged under 15 (Gallagher, 2019), although this is not to assume that children's labour per se is a problem (e.g. Wyness, 2013; Jonah and Abebe, 2019). Gallagher goes on to report exploratory qualitative research that focuses on the conditions in which children are working, as they physically extract the precious cobalt from rock, wash and transport it. The research identifies at least two sets of harm experienced by children: poor and unregulated working conditions, including malnutrition, carrying excessive loads and abuse from adult supervisors; and the direct negative health effects associated with exposure to cobalt given a lack of protective clothing.

Interestingly, media representations *of* these entanglements of rock, media and childhood are divided – some focusing on the risks faced by children, others on the benefits of such work, with some middle-class children choosing to work in the mines to supplement their families' incomes for school fees. Yet Gallagher's broader point is that, often, the component parts of these entanglements are viewed separately, whereas (as I read it) what is needed is the kind of machinic, post-human media ecology that Bryant articulates. This "demonstrates the potential of analyses that join up different aspects of technology – social, technical, (geo)physical. This sort of multi-layered analysis takes seriously the task of thinking more relationally about childhood and its materialities" (Gallagher, 2019: unpaginated) – and in ways that extend beyond even the more radical forms of new materialist thinking about childhoods explored in Chapter 1. Moreover, to understand the "unit operations" (Bogost, 2006: 1) of the geology of media requires upscaling: "articulating how the micro-politics of child mining is just one part of a massive planetary assemblage that also involves Chinese ore buyers, refining processes, shipping, tech manufacturing factories, plastics, handheld devices, consumers, wireless radio signals, and so on" (Gallagher, 2019: unpaginated). Ignorance of these processes – especially, as Gallagher points out, amongst the adults and children who use devices containing cobalt processed by Congolese children – is a form of Marxian alienation. This then is a powerful argument to go beyond analyses of use, which, whether knowingly or not, efface or hide the *power* of machines through a Minority World optic. Thus, Gallagher (2019: unpaginated) concludes:

> [a]gainst the popular view that media are destroying children's relations with the natural world [. . .] media are systems through which children are in fact entangled with the earth. In the case of cobalt and other media minerals, problematic but potentially provocative relations [. . .] are hidden by the superficiality of utilization. If we want our media to help us formulate ecological politics for the Anthropocene, it might be worth finding ways to push beyond a focus on utilisation, get behind the screens, and dig into the social-technical-physical relations that constitute these technologies.

In my reading, Gallagher's work is exemplary of the wider point I am trying to make in this chapter, and indeed in this book. That is, he foresees a social-technical-physical analysis of media and machines that at once extends beyond children yet is – perhaps all the more so – concerned *with* children and with the *multiple, differentiated childhoods* that our obsession with (or cruel optimism for) digital media is constituting (compare Berlant, 2011).

Hyperobjects

Bryant's discussions of flow and power – and Gallagher's of the planetary scale of geologies of media – link nicely to a second style for thinking materials and childhoods: hyperobjects. Morton's (2010, 2013) theorisation of hyperobjects shares much with Bryant's in that it, too, is a brand of object-oriented ontology. Thus, it shares premises with Bryant's work. However, Morton's thinking is couched in a problem: a problem caused by the exertion of particular kinds of power (as Bryant has the term), by some very particular kinds of objects. The problem – which resonates with the discussion of *resource-power* in the previous chapter – is an ecological one. It is the ecological predicament of the world (more on that term in a moment) in what some commentators term the 'Anthropocene'. Often specified in Morton's work by a concern with global warming, this ecological predicament might also manifest in the form of irradiation by nuclear materials, swirling piles of plastic waste or the ubiquity and 'stickiness' of oil (Negarestani, 2008).

Whether or not one agrees with the designation 'Anthropocene' (e.g. Haraway, 2016), Morton's starting point is with a concern with earth's present ecological condition. For him, it is no longer possible to talk of 'the earth' – or, rather, 'the world' – because what he terms 'hyperobjects' have brought about the end of the concept 'world'. This is largely because a subset of the species 'human' has spent centuries trying to distinguish itself from the rest of life on earth precisely through the designation of a concept of 'world' that sits exterior to the human, can be known as such and can be subject to exploitation and mastery. He is clear that, of course, planet earth exists. Yet – as many other subsets of the human species have known for a very long time – the idea of *the world*, as it has been deployed in certain, hegemonic, capitalist and Modern styles of thinking, is no longer tenable. It is – to anticipate an argument I make at the very end of the book – effectively *dead*.

In Morton's logic, the end of the world-as-concept is a function of the operation of hyperobjects. Much like other theorisations of the Anthropocene (such as Yusoff's conception of geo-power, which I discussed in the previous chapter), this is because hyperobjects have inverted our thinking styles. No longer are our thoughts or feelings our own – if they ever truly were – but they are 'footprints of hyperobjects' (Morton, 2013: 5); there is no (world) outside; we have been scooped out by and inextricably enmeshed within these hyperobjects. In my reading, this realisation ushers a double manoeuvre. On the one hand, in Bryant's terms, hyperobjects are particularly and peculiarly insistent, as I shall explain shortly. They insist that we turn the lens inwards, on our constitution by matters so large (or so tiny) that we rarely notice them directly. But, rather than the world only existing *for us* (as traditional phenomenologies and correlationist thinking would have it), these hyperobjects *become us* – and we become them – in ways that cannot be explained by phenomenological concepts such as Heidegger's *Dasein*, which insists upon the specialness of (particular) humans in their ability to know the world. This is because, on the other hand, and chiming with a key tenet of many forms of OOO, there is a gap between hyperobjects and our ability to perceive them directly. As Morton argues, "[t]he gap between phenomenon and thing yawns open", because "'hyperobjects are not simply mental (or otherwise ideal) constructs but are real entities whose primordial reality is withdrawn from humans" (Morton, 2013: 12 & 15). Hyperobjects are all part of the displacing of human life – the decentring that is such a concern for this book – that forces us to recognise that neither are we 'centre stage' but nor can we leave the stage entirely. As per the central arguments of Harman's (2010a) approach to objects (outlined in Chapter 1), objects are both partially withdrawn from one another, and object-relations – to the extent that these can occur – may occur beyond human perception, intentionality or agency (also Bogost, 2012).

Morton characterises humans' relationship with hyperobjects in the Anthropocene as "a new phase of *hypocrisy, weakness,* and *lameness*" (Morton, 2013: 3), a great humiliation that brings humans crashing back to earth. This is in fact a double movement, and a twofold optic: "a double denial of human supremacy" akin to film director Alfred Hitchcock's fabled *pull focus*, with a dose of the Freudian uncanny. Morton (2013: 19) writes: "By simultaneously zooming and pulling away, we appear to be in the same place, yet the place seems to distort beyond our control." Thus, hyperobjects' insistence operates via a twofold acknowledgement: an awareness of the gap between objects and our ability to perceive them; and a recognition of our ineluctable entanglement and co-constitution with, in and as hyperobjects. This acknowledgement is methodological, political and ethical, as much as conceptual (compare Bennett, 2010). However, it extends and distends notions of an ethics of 'common worlds' (e.g. Taylor and Pacini-Ketchabaw, 2018) because the more-than-human objects under question are not so readily identifiable, haeccetic or bounded as animals or toys. Most importantly for the purposes of this book, the 'pull focus' enacted by hyperobjects is an extension to my argument in the previous chapter for what I termed, after Tsing (2015), 'arts of (not) noticing', and I therefore return to the idea of the 'pull focus' throughout the book.

Having established something of the force or insistence of hyperobjects – why they matter – it is now apposite to consider what hyperobjects *are* and the properties through which they operate. Hyperobjects are "massively distributed in time and space relative to humans" (Morton, 2013: 1). They are hyper-relative to both humans and other entities, and may or may not have been created by humans. Examples cited by Morton at the opening of his 2013 book include black holes, oil fields, the Florida Everglades, Styrofoam cups and plastic bags. The waste at the Guaratinguetá eco-park – as part of an extensive, if non-contiguous mass of waste whose tentacles and blotches taint many parts of our planet – could be another example. I will cite numerous other examples of hyperobjects throughout this book, with plastics (amongst others) being particularly prominent. Indeed, plastics – whether found at an eco-park, an apartment block, in an oceanic vortex or on an online selling site – constitute hyperobjects par excellence, as they are ready reminders of the twofold logic of hyperobjects. Matters like plastics – such as the plastics in the example with which I end this chapter – are able to circulate globally given how readily they can be dispersed and ingested. Taken as an 'object' – perhaps, in relation to Bryant's (2014) evocation of machines, or Bogost's (2012: 23) of 'units', a less elegant term in this particular case – plastics are massively distributed across and in the earth. Like other matters that exist at the nano- and micro-particular scale, we know they exist – we make some of them, we put others in children's sun cream and others still in tomato ketchup – yet we cannot readily perceive their simultaneous minuteness and enormity. Meanwhile, as a result of their operating at both ends of the (earthly) scalar spectrum, we are becoming increasingly entangled with plastics in ways that are only now starting to become clear. This is part of the argument I will pursue in Chapter 8, when I look at (or, rather, start with) plastics in as much detail as my own faculties allow.

With these examples in mind, it becomes possible to sketch out some of the properties of hyperobjects, which both overlap with and are distinct from those of machines. An important starting point beyond the sheer scale of hyperobjects is their *viscosity*. Think of the nano-containing sun cream, which is programmed to adhere to the skin. Or, as Morton (2013) urges, what happens if we flush away a wadded tissue covered in a child's vomit: it does not go 'away' after going around the U-bend – it goes to the Pacific ocean or to some kind of treatment facility. It does not disappear. Another good example "of viscosity would be radioactive materials. The more you try to get rid of them, the more you realize you can't get rid of them" (Morton, 2013: 36). Thus, "[v]iscosity is a feature of the way in which time emanates from objects, rather than being a continuum in which they float". Things *stick* to hyperobjects, and/or hyperobjects coat things (including us, other things, even the earth). They do so for different periods of time, hanging around with different speeds and slownesses, sticking to us for different periods of time, and hence they emanate time. Plastics decompose at different rates from wadded tissues, and again from radioactive materials.

An extension of the hyper-scaling of hyperobjects is their *non-locality*. Here, there is an overlap with Bryant's (2014) notion of 'local manifestations' of machines, which I discussed above. Morton's analysis of non-locality is replete with quantum theory, the details of which extend well beyond the use I want to make of his work. However, a useful addendum to Bryant is the sense in which hyperobjects – because of their size – involve "[a]ction at a distance" (Morton, 2013: 39). Key here is that local manifestations of objects are not *the* object. As Morton argues, we can't directly *sense* global warming – the sun burning our heads or a raindrop may well be a local manifestation, but it is a "false immediacy" (Morton, 2013: 48). Although we can't necessarily *see* or *feel* global warming as a whole – we cannot grasp it as an object – we can piece together evidence of its existence through what I later term *instrumental interfaces*: a global temperature chart here; testimonies of living through extreme weather events there; evidence of species decline here; the ancient art of phenology, telling us of changes to seasonal timings there. This is because although many OOO theorists steer a path away from relations or networks (as per Latour), they do hold onto some sense in which objects might (partly) relate. For Morton (2013), hyperobjects are *interobjective* – they are detectable through relationships with objects. For Bryant (2014), machines are insistent and, because they are malleable, are subject to change through interactions with other machines. For Bogost (2012: 25), the "word *operation* [describes] how units behave and interact" in order to produce transformations. These are all ways of talking about how objects exist in themselves, but also (perhaps temporarily) constitute parts of larger objects or are constituted by smaller objects. Thus, especially with some distance, the hyperobject 'global warming' is configured through complex interactions of water, gases, rocks, wind and more; with some distance – especially with time – the enormity and complexity of those configurations comes partially into view. Indeed, the different modes we have for perceiving global warming – such as those listed above – are in themselves configurations, as much as they are configured-with global warming.

Furthermore, and as the example of the sheer *time* it has taken to gain *some* sense of global warming attests, hyperobjects create *temporal undulations*. Because they are so enormous, we cannot see their start and end points, whether in space and/or in time. This massive distribution baffles humans, because it is hyper to our habitual spheres of perception. For instance – and sticking with the stickiness of global warming – Morton offers the example of starting a car engine. When we do so, and when we also stop to try to think the hyperobject named 'oil', we experience the uncanny recognition that "liquefied dinosaur bones burst into flame" (Morton, 2013: 58). Careering into the distant future rather than the very distant past, we acknowledge that "7 per cent of global warming effects will still be occurring one hundred thousand years from now as igneous rocks slowly absorb the last of the greenhouse gases" – timescales that are, quite literally, "petrifying" (Morton, 2013: 59) – as well as they are humiliating, as I have already argued. Critically, and to return to the example of plastics with which I end this chapter, this is a recognition that there are hyperobjects, and local manifestations thereof, which we may see or hold today, but which will exist well into the future because they take so long to deteriorate. With plastics or concrete, we see into the future – a future not here, not over there, but nevertheless "a real entity in the real universe" (Morton, 2013: 67) and one that could be terrifying.

The final term in Morton's lexicon is *phasing*. Like the sun, hyperobjects phase in and out of the human world: "hyperobjects are *phased*: they occupy a high-dimensional *phase space* that makes them impossible to see as a whole on a regular three-dimensional human-scale basis" (Morton, 2013: 70). A mountain might be a similar example: as something that cannot be experienced in its totality and as something that becomes in some senses more evident with distance (although distance does not equate to objectivity). The notion of phasing brings together the different conceptual elements of hyperobjects. Because hyperobjects are transdimensional and non-local, we only see parts at a time and – in the terms of an oft-repeated trope of OOO – certain elements of those objects are withdrawn (Harman, 2010b). Thus, to return to Morton's key example, we could ask: what would global warming look like if we could become big enough to see it? Or, what would the sum total of all plastics on earth look like if we could both become small enough, and simultaneously big enough, to see them? Of course, these are hypothetical questions although, as I argue in Chapter 8 in particular, this does not mean that we should not attempt to broach them in some creative and 'radically' interdisciplinary ways. Indeed, Morton himself advocates for such a move, considering how combining analyses from the arts, cinema and climate science might generate at least affordances of global warming.

The phasing effects and affects of hyperobjects mean that they waft in and out of view, in and out of our fields of attention. Thus, "[m]y attention space focuses on global warming for a few seconds each day before returning to other matters" (Morton, 2013: 74) – with the term 'global warming' interchangeable with any other hyperobject. Or, "[p]hasing means to approach, then diminish, from a certain fullness" (Morton, 2013: 74). The phasings of hyperobjects – and our attempts to acknowledge them – are together bound up in a by-now familiar double movement of the pull focus: zooming in and out of focus, simultaneously decentring and recentring, evoking arts of (not) noticing.

My question – which I begin to broach in the chapter's conclusion – is twofold. Firstly, what happens when we attempt to take the point of view of hyperobjects rather than humans; or, rather, what happens if – taking the speculative leap of faith that OOO asks us to – we attempt to take the perspective of non-human traces of hyperobjects, seeking to map their phasing? To do so would, in part, be to make visible networks or "imbroglios" of stuff (Latour, 2005: 46), or to "follow the thing" (Cook, 2004). Secondly, what happens when we ask: what are the hyperobjects that are (perhaps particularly) constitutive of, or matter to, *child-hoods*? Or, again: *where are children*, precisely?

Plastic (childhoods): off the island of Roatán, 40 miles north of Honduras, in the Caribbean Sea

In order to begin answering the questions I posed at the end of the previous section, and in order to provisionally tie together my discussion of machines and hyperobjects, I want to turn to the example of plastics. As I write this, concern in the United Kingdom about the environmental impacts of plastics has rapidly intensified following the broadcast of an episode of the natural history TV series *Blue Planet II* in December 2017. The programme showed some of the effects of plastics on the life (and death) of the world's oceans. What is so striking about this plastic moment is the sheer *scale* of the debate and of the very material processes in which plastics are entwined. Generically, plastics constitute a hyperobject, as Morton conceives them. Moreover, certain subspecies of plastic – like Styrofoam cups, but also those associated with children, like nappies (diapers) and dolls – could also constitute hyperobjects, given their virtual ubiquity.

Cut to a video posted on *Boing Boing* (2017). At least in the United Kingdom, we are familiar with images like this – a still taken from a video off the island of Roátan, in the Caribbean, 40 miles north of Honduras. The amateur film documents a tourist boat ride and an encounter with a huge raft of detritus (a 'garbage patch' – see Huang, 2017). It is one of many oceanic trash 'vortices', the most famous and largest being in the Pacific Ocean: massive rafts of assembled detritus, swirling entangling, and – like the stuff in the abandoned apartment block in Luz – decaying with different speeds and slownesses. They are composed not only of plastics but of woods, metals, vegetation, feathers and, no doubt, an array of marine life, both living and dead.

As the bemused voices in the background of this video (apparently taken on a tourist boat ride) list off the objects they see, *plastics* are prominent. Moreover, although this listing seems arbitrary, quite significantly, embedded in this listing are several 'ordinary' plastic objects constitutive of especially Western childhoods. For me, their listing is reminiscent of Bruno Latour's (2005) evocation of the uncanniness of listing apparently ordinary objects. As Latour argues, to apprehend the most quotidian of objects from an estranged vantage point (an idyllic Caribbean boat-ride that takes on a somewhat monstrous turn), and to list them off (as we hear a female American voice calmly pointing out sandals, dolls, toy cars, bottles – some of the very same kinds of stuff at the Luz apartment), is to make it matter, again, or differently. Latour says:

> even the most routine, traditional, and silent implements stop being taken for granted when they are approached by users rendered ignorant and clumsy by *distance* – distance in time as in archaeology, distance in space as in ethnology, distance in skills as in learning.
>
> (Latour, 2005: 80)

Although perhaps not intentionally, Morton's treatment of the distance and uncanniness heralded by hyperobjects seems to resonate with Latour's consideration of distance. Like the boat-ride, and like our (in the United Kingdom) sudden attentiveness to plastics, there is also an obvious *phasing* going on here. Plastics are 'on the agenda', to the extent that some commentators are concerned that plastics are a "dangerous distraction" from the effects of other hyperobjects, and especially climate change (BBC, 2018a: unpaginated). Indeed, there are multiple phasings happening here: the phasing of plastics into and out of the boat-ride (on another day, ocean currents could have been such that the vortex might not have crossed the route); the phasing of plastics, in and as media (like the video, or *Blue Planet II*, or the coverage that has ensued), as an object of public concern; and, let us not forget, the brief, bare phasing in and out of childhood (objects) in the video commentary (reminiscent of the apparently arbitrary listing of childhood in Bryant's [2014] work, as discussed above).

To *recognise* the objects in the video, and objects like them, as 'hyperobjects' is – for me – an interesting exercise. Yet my consequent and more pressing concern is with the *power* of these hyperobjects. Indeed, connecting Morton (2013) and Bryant (2014), I want to ask how the powers of hyperobjects wax and wane – how they are phased – as they are coupled or uncoupled with other (hyper) objects, such as a boat-ride, media hysteria about plastics, or childhood (objects). It is perhaps here that, with my enduring interest in childhood, I differ from the flat ontology proposed by some OOO theorists, amongst others (Bogost, 2012; Schatzki, 2016). Or, perhaps, certain brands of OOO deploy flat(-ish) ontologies in the service of analyses of power, if the term 'power' is unbridled solely from questions of *human* power. As Bryant (2014: 52–53, my emphasis) puts it, in a passage that could very well be referring to hyperobjects, power is a question of *attention*:

> [i]t is a machine that is capable of resonating in a variety of ways given the historical and cultural milieus that it encounters. It's as if there is a certain vagueness, a certain floating nature, that characterizes these works allowing them to maximally traverse culture and history. With pluripotent works such as this, we get a reciprocal determination. They both act on their historical and cultural milieu and are acted upon by their historical and cultural milieu. The milieu actualizes the work in a particular way [. . . but] also organizes the historical and cultural milieu in a particular way leading us to *attend* to certain cultural phenomena as significant while ignoring others.

Once again, then, this is a question of phasing and of attuning to arts of (not) noticing. However, if this exercise might appear to be abstract and arbitrary, then Bryant's (2014) invocation to what 'leads us to *attend*' is an important one. Whether we agree with them or not, and whether they are anthropocentric in nature (and that is something we should perhaps always keep an eye on), the phasing of some hyperobjects is such that they demand our attention. Or, they *matter* (Horton and Kraftl, 2006a), in different ways, in different times and places. As Gallagher (2019) shows, cobalt articulates particular kinds of social-physical-technological mattering. In different but related ways, plastics matter: to the earth; to the marine species whose lives they endanger; to what Bryant terms the organisation of our 'historical and cultural milieu'; and, as I shall argue in later chapters, to the ways in which *children* and *childhoods* become entangled with plastics.

Conclusion: thinking childhood through machines and hyperobjects (and dolls, cobalt and plastic)

Childhoods have been largely out of focus in this chapter, as I have been concerned with sketching out some of the key conceptual tools that nevertheless are vital to thinking and doing, after childhood. At times, childhoods, children, and material or affective traces of both have appeared – sometimes obliquely; sometimes ambivalently; sometimes arbitrarily – and/or sometimes childhoods have phased into view in the sharpest and most concerning ways, even if the voices or experiences of individual children have been an absent presence. I used Morton, Bryant and a range of other sources – some self-identifying as object-oriented ontologists, others who think materialities in different ways – to develop and extend the argument that, in order to think and do, after childhood, children must be decentred and recentred from our analyses. In general terms, these arts of (not) noticing are a deepening of the attention to resource-power – and the positioning of childhoods therein – in the previous chapter. In more specific terms, I have sought to introduce a range of pressing challenges facing some of the world's children – some familiar to childhood studies scholars, others perhaps less so because they involve *not* always starting the narrative with children. These challenges ranged from urban 'vacancy' and poverty in São Paulo, to cobalt mining in the Democratic Republic of Congo, to the media ecologies of plastic pollution in the world's oceans. I am arguing, then, that the tools developed in this chapter might not only be helpful if we are interested conceptually in 'machines', 'hyperobjects' and their entanglements with childhoods. Beyond that, as Gallagher (2019) argues so powerfully, they could help activate an awareness and critical analysis of – without wishing to fall into the same technocentric, heroic conceits of Anthropocene discourse (Taylor, 2019) – potential modes to address those challenges.

Rather than repeat the key features of machinic and hyperobjectified thinking here, I close this chapter by briefly outlining some potential questions and tasks for thinking and doing, after childhood. Taken together with the rest of the chapter, these both extend and specify the introductory discussion of OOO and materiality in Chapter 2. Firstly, thinking with machines and hyperobjects might guide childhood studies scholars to witness, but also to go beyond, an awareness of the "reciprocal determination" of childhoods with non-human materialities – something that has been evident at least since Prout's (2005) *Future of Childhood*, if not before. I contend that the conceptual languages of machines and hyperobjects extend our field of vision – to different types, scales and temporalities of the material, including stuff like plastics (also Parikka, 2013; Horton and Kraftl, 2018). I also posit, though, that we must "contend with the power characterizing the being of these flows [. . . since] machines are not sovereigns of the flow that pass through them" (Bryant, 2014: 50). Drawing on Bennett (2010), Bryant cites the example of how machines (young bodies) maybe modified by other machines that pass through: how omega-3 fatty acids, given as supplements to children with special educational needs and young prisoners, might improve their attention spans, whereas other fats may have different physiological effects on those bodies (for instance, making them larger: Bennett, 2010: 41). The implications of such machinations are manifold: from critical perspectives on the ethical implications of intervening into and monitoring children's bodies (e.g. Evans, 2010; Gagen, 2015), to questions about the evolutionary capacity of human bodies to cope with generational shifts in diet or lifestyle (Chapter 6), to a need to analyse which humans get to exert 'the power characterizing the being of these flows' (Bryant, 2014: *op cit.*). In any of these cases – and others – machines and hyperobjects afford a sense of extension: of *childhoods* extended and of our potential *analyses* thereof extended.

Secondly, an attunement to hyperobjects in particular does more than bring into view fields of diverse non-human *agents* in the constitution of childhoods – indeed, this recognition now structures much empirical and conceptual work in childhood studies. Rather, it also requires a sensitisation to more-than-human spatial and temporal *scales* that have hitherto been virtually ignored by childhood studies scholars and, as Bogost (2012) argues, by thinkers of materiality more generally. As I will show in later chapters, Bogost charges that many brands of material-thinking – including those seeking to witness the actancy of the more-than-human, such as Actor Network Theory (ANTs) – remain in some ways human-focused. This is because they do not attempt to understand what *it is like* to be a non-human object, or machine, or unit. Whilst in a book on childhood I can only take this logic so far, it nevertheless opens out possibilities for a parallel manoeuvre: to attempt to understand what (for instance) non-human generations *are like*, as they extend, interweave through and endure beyond human generations. Morton's (2013) concerns with the 'petrifying' timescales of global warming are a case in point.

Finally, I have begun in this chapter to ask how – heeding especially Morton's warnings about the overwhelming elusiveness of hyperobjects, and Harman's about objects more generally – it might be possible to nevertheless find modes for visualising entanglements of children with machines and hyperobjects. And, in turn, it might be possible – indeed, it is, really, a pressing task, as Gallagher notes – to articulate how those entanglements are riven with power, differentiating childhoods in ways that maybe complex and contradictory. I do not purport to have the answer, and my steps here are tentative. One possible response, with which I have experimented so far in this book, is with the telling of stories – and here I owe a debt to others who advocate for such a strategy in the face of grand Anthropocene narratives (e.g. Taylor, 2019). Another possible response is to think about what matter and media make 'visible' – a task I broach in the next chapter. And another possible response, with which I experiment in Chapter 8 in particular, is to carefully consider more 'radical' forms of interdisciplinarity – weaving stories from environmental nanoscience *into* the kinds of stories that I am certainly more comfortably telling about childhoods – without assuming that scientific knowledges offer a way out of the trouble (Haraway, 2016).

5   Media

Visibility, circulation and some stuff about childhoods

Aleppo, Syria, August 2016

On 1 August 2016, the BBC reported that children in Aleppo were burning tyres (BBC, 2016). The resultant smoke would create a 'smoke curtain' that would reduce visibility for warplanes that had been bombing the city, therefore creating a no-fly zone. In turn, the city's residents hoped that this action would reduce bombing of the already beleaguered city. At the time, many commentators considered that this was probably the only effective act of resistance available to children; indeed, whilst children's actions may have reduced the bombing in the short term, the civil war in Syria continues.

The news article about the children's actions was very short – in fact, much of the piece was taken up with images circulating on social media through the hashtags #AngerforAleppo and #AleppoUnderSiege, and with discussion of a Russian helicopter shot down by rebels a few days earlier. One particularly striking image being shared on Twitter was a cartoon-style, black-and-white, silhouetted image of a single young child blowing smoke out of a car tyre on a stick (made to look like a child blowing bubbles) whilst a fighter jet flies overhead and the crumbling remains of the city burn around them. The proportion of the child's head to their body shows someone probably no older than 6, even though the children appearing in photos related to the tyre-burning look much older and are all in large groups rather than alone. Meanwhile, one of the factions fighting in this complex and multi-sided conflict – the Free Syrian Army – tweeted a bi-tonal, black-and-red image containing the hashtag #AngerforAleppo, with a stylised burning tyre above it.

Trawling in detail through the two hashtags above on Twitter, the content is varied. It does not focus only on the children burning tyres but upon political slogans, videos of the bombing, adverts by NGOs and charities indicating what people in other countries can do to help, and far more besides. Children, it appears, are not the sole focus of the (overwhelmingly) adult Twitter users posting text and images. When they are, the messages are short and to the point; they are often accompanied by one or other of the images described above; and the words are replete with a sense of despair or lost hope for these children. For instance:

> The children of Aleppo create their own #NoFlyZone coz world has let them down

Children in #Aleppo are burning tires to create a #NoBombZone. We must help them #StopTheBombs. #AngerForAleppo
  (accompanied by a one-minute video interleaving shots of children rolling and setting fire to tyres, and bombs hitting the city)

Imagine yourself come back home, and found your kids dead Is the #Anger-ForAleppo enough?? Do something

#SyrianChildren in #Aleppo r burning tires 2 create a #NoBombZone #Stop-TheBombs #AngerForAleppo #Syria #World shame!

Quantitatively speaking, tweets referencing children make up only around a quarter of the content. Yet they were a part – *the key part* identified by the BBC and many other international news agencies – of a deliberate attempt to create a Twitter surge or 'storm' around the plight of Aleppo's residents, using the #AngerFor Aleppo hashtag. Whether or not the 'storm' was *effective* in any way, it is virtually impossible to tell. Yet, in a manoeuvre that is all-to-familiar to many childhood scholars, the agency of children – which might be theorised, as per Chapter 3, as a form of 'resource-power' – is re-positioned and, crucially, re-presented such that the (cartoon) figure of the child is deployed to offer a poignant, emotive or what Wells (2013) terms 'melodramatic' reminder of lost hope.

The Syrian civil war has been one of the most devastating of the twenty-first century. I do not intend to offer any further analysis of the war itself. Nor will I analyse any further the content of the tweets and images circulating about Aleppo's children. Nor can I consider the social and political agency of the children in this or any other context after the Arab Spring of 2011, or, indeed, their own social media usage during those struggles (Jeffrey, 2013). Rather, I want to broach two different questions, implicit above. What is the nature of the connection between the children's actions and the images and text circulating on Twitter? And what (and who) is rendered (in)visible?

In posing these questions, the intention is not to evade the challenges of analysing the Syrian war (although I am woefully under-qualified to do so). Nor is it, conversely, to raise this knotty, difficult and troubling conflict as a tokenistic exemplar for an aloof conceptual argument – although, as with the case of Luz in the previous chapter, I accept that I run this risk. Rather, in responding to the above questions, I want to interrogate two further problems and the possible connections between them – problems that, in my view, have not yet been subject to sufficient analysis. In both cases, the problem is with the circulation of *stuff about childhoods* – and, indeed, the circulation of childhoods themselves. On the one hand, I am curious about how it might be possible to analyse, visualise, present and subject to critical scrutiny the circulation of images and discourses about childhood through digital (and social) media such as Twitter. As Rose (2016) argues, digital 'cultural objects' are not necessarily different from others that have preceded them, and the same argument might be made about the media that contain what childhood scholars have for years termed the 'social construction of childhood'. Yet, in parallel with recent arguments about technology's role in the biopoliticisation of childhoods (Lee and Motzkau, 2011), circulations of childhood via digital media do differ. They differ in their speed, quantity and intensity; and, if not in their content, in what Rose (2016) characterises as a move from representation to 'production', as interfaces with and networks of digital media involve proliferating human and non-human actors, software and hardware, including Artificial Intelligence (AI) and algorithms. Writing on the rise of digital platforms in the education sector, Williamson (2017: 62) puts it thus: "[i]nfrastructuralized platforms are, then, becoming as integrated into contemporary society as existing infrastructural networks of transport, electric utility, broadcast, print media, and telecommunications". As Williamson and others writing on the 'datafication' of childhood (also Willson, 2018) attest, and as I explain in more detail later in this chapter, this means that the sheer amount of stuff *circulating* about individual children, and about childhoods generally, is becoming both increasingly intensified and progressively normalised.

On the other hand, and returning to the trope of materiality, matter and object-hood that also runs through this book, and which particularly detained me in the previous chapter, the tyre-burning children of Aleppo raise a series of concerns about the *mattering* of childhood. Without wishing to tokenise the example, it nev-ertheless speaks powerfully about how particular, partial, material encounters (as Harman has it) or media-ted operations (in the multiple ways Bryant understands them) are not merely esoteric, abstract concerns. Interfaces between objects and objects (between tyres, smoke and aeroplanes) and between objects and humans (children, aircraft pilots and the unseen leaders issuing orders to bomb Aleppo) *matter*, profoundly, in situ. Such encounters would look very different in other con-texts, both in the interfaces between objects and their mediation. Thinking simply of the encounter between aeroplanes and smoke, other equally challenging local manifestations have occurred that looked very different, including the September 11 attack on the Twin Towers in Manhattan; or the 2010 ash cloud issued by the Eyjaf-jallökull eruption in Iceland that saw flights grounded or diverted; or, on the day I drafted this chapter, news and social media coverage of a British Airways flight to Valencia in Spain whose cabin filled with smoke minutes before landing. The point is that objects only ever encounter one another *partially* and *differentially*, with or without significant input from humans, and with more or less intended effects. But to add to this common refrain of OOO and speculative realism, when humans *are* involved somehow, these encounters not only happen partially and differentially but they *matter* partially and differentially. They matter, whether in the sense of the inconvenience of a flight delayed by an aberrant Icelandic volcano, or, in one of the most dangerous and contested cities in which a child could grow up, and in relation to a seemingly intractable challenge that involves not only a multifaceted conflict but attendant problems associated with an accompanying drought, a spike in food prices, forced migration and the destruction of physical infrastructures in Aleppo and other Syrian cities. Thus, whether (ethically and politically) events such as these appear ephemeral and 'silly', or violent and 'traumatic', depends very much on the specific timings and spacings of interfaces between humans, media and materialities like smoke (see Chapter 9 for more on these interfaces).

Meanwhile, and perhaps ironically, the children of Aleppo chose a hyperobject par excellence – the car tyre, that ubiquitous materialisation and symbol of late capitalism, and of the 'independence' promised by the latter. Similarly ironically, in choosing to burn the tyres, they (or rather, the particular interaction between fire and tyre) released micro-particles into the atmosphere. As those of us who drive do also, albeit less visibly and less acutely, they contributed to the circulation of a range of pollutants in the atmosphere, thus helping to ensure the tyre's status as hyperobject. And more ironically still, and if this reference to air pollution seems like a tenuous link, in a move that was somehow darkly humorous, satirical and poignant, Aleppo residents offered on social media 'apologies' on behalf of the children for contributing to climate change and atmospheric pollution, in a video now viewed over 16,000 times and posted with the #AngerforAleppo hashtag.

Thus, with these two problems in mind, I want to ask again: what is the nature of the connection between the children's actions and the images and text circulating on Twitter; and what (and who) is rendered (in)invisible? This is, again, a question about *where children are, precisely*, in these kinds of interfaces. It is my observation that, as I will argue in the next section of the chapter, despite (some of) the proponents of digital media scholarship arguing for the enmeshing of materialities with digitalities, the concerns of that scholarship remain largely distinct from matters like tyre-burning. Meanwhile, and in the same vein, aside from a few notable exceptions (e.g. Smith and Dunkley, 2018; Land et al., 2019), childhood scholarship that *is* concerned with more material matters – particularly new materialist and post-humanist work – tends not to engage with digital cultures-technologies. And, in my view, neither adequately grapples with either the circulation of stuff about childhood (and here I mean more than data and 'datafication') or the nature of the relationship between, for instance, tyre-burning and the thousands of posts with the #AngerForAleppo hashtag.

It is here that I argue the work of Levi Bryant (2014), and his theorisation of media and machines – introduced in the previous chapter – could offer a useful point of departure. Thus, after briefly overviewing the now vast literatures on digital/social media and youth, I articulate how an expanded conception of 'media' could help explain not only the multiple (and multiply mediated) matterings taking place in Aleppo but a range of other imbrications of digitality with childhood. Because a common refrain that develops through this chapter is the notion of *circulation*, Bryant is joined by a number of other theorists who discuss the movements, networks, journeys and flows of digital media, as well as *interfaces* between media, humans and other objects as particular kinds of unit operations. Empirically, I start with some original analyses of Twitter and an online selling platform data undertaken during the Plastic Childhoods project, before moving on to juxtapose a range of examples, taken from around the world, of the use of AI, Big Data and algorithms in and around children's lives. Herein, I develop and expand some of the conceptual arguments introduced in the previous chapter, whilst anticipating a number of themes that I pick up in more detail later in the book – not least around concepts of generations, energy and juxtaposition.

Interfaces between children, young people and (digital) media: rendering objects (and subjects) more or less visible?

> [T]he materiality of the medium, its material properties and powers, substantially modify human activities and relations in ways that [often] *outpace* the *content* of the medium.
>
> (Bryant, 2014: 32, my emphasis)

> [A]s educators and parents, we should be thinking about the implications for digital play and learning when the technology is no longer *visible*, because these changes are happening fast.
>
> (Plowman, 2019: 33, my emphasis)

The previous chapters have – through an introduction to new materialist, nexus, post-human and speculative-realist thought – sought to compile a range of ways for theorising materialities, after childhood. There, a principal concern was that, in the absence of an attention to 'things', "an entire domain of power became *invisible*, and as a result we lost all sorts of opportunities for strategic intervention in producing emancipatory change" (Bryant, 2014: 3; my emphasis). In this chapter, taking the notion of the 'pull focus' in a rather different direction, I want to ask – drawing on the two quotations that begin this section – what is made more or less *visible* by both the *speed* and (im)material properties of the digital. This means somewhat of a juxtaposition and a risk of a conceptual rupture, exemplified by my questions about the relationship between tyre-burning and Twitter in the Aleppo example. But this juxtaposition is in keeping with the sense in which I am developing arts of (not) noticing childhood. Initially, I offer a brief review of literatures on digital media, childhood and youth, drawing out a series of implications and questions for the (deliberately broadly conceived notion of) 'visibility' that acts as a navigational aid throughout this chapter.

Mirroring the rise of digital media over the past two decades, there has been a surge of interest in children and digital media in academic research (for detailed reviews, see Ergler et al., 2016; Blum-Ross et al., 2018; Goodyear and Armour, 2018). As many commentators note – and have noted for nearly two decades now – academic debates about children and digital media tend to grapple with, on the one hand, often sensationalised debates about the risks and threats of (especially) the internet and, on the other, the quasi-utopian or 'revolutionary' opportunities for children's emancipation, play and learning (e.g. Holloway and Valentine, 2001; Selwyn, 2003; Bond, 2014; Thompson and Sellar, 2018). Whilst a considerable proportion of work focuses on the former – on the risks and threats to children's development, health, well-being and socialisation – there are also plenty of examples of balanced, critical analyses that have, for instance, highlighted the benefits of the internet and social media for young people's improved access to information about health and well-being (an exemplary case being Goodyear and Armour, 2018).

A key thematic and conceptual consideration has been the role of digital media in constituting children and young people's identities. For instance, both Boyd

(2014) and Ito et al. (2010) use the notion of 'networked publics' to examine digital media's capacities for identity- and group-formation. Whilst the *processes* sitting behind identity formation may not differ hugely from those afforded by previous generations of media, one important strand to pull out of this work is the increased *visibility* that ever-faster and more intensive forms of networking bring. In other words, as Ringrose et al. (2013) show in their important work on sexting (the sharing of text and other messages with sexualised content), digital media offer enormous capacities for the (re)presentation of self-images, often edited and carefully curated in ways that are partial, if not quasi-fictional. A second important strand from this work is attention to what Ringrose et al. (2013) term the 'after-lives' of such forms of visualisation – the ways in which particular images circulate through and beyond friendship networks, collect 'likes' or, even, go viral. Inevitably, these image-heavy cultures of visualisation have attracted criticisms and a significant amount of empirical research about their impacts upon young people. Although many findings remain somewhat contested – especially around the effects upon children's neural development (Choudhury and McKinney, 2013; Loh and Kanai, 2016) – reported effects include a deterioration of attention spans, thinking processes, communication skills and social-connectedness (e.g. Turkle, 2017).

Although arguably most significant for crafting a careful and critical case for a rights-based framework for analysing and governing children's use of digital media, Sonia Livingstone's work on audiences also offers other insights into questions of visibility (Livingstone and Third, 2017; Livingstone et al., 2017; Livingstone, 2019). Livingstone (2019) argues that one implication of theories of mediatisation – which emphasise how everything and everyone is becoming mediatised, ever-faster – is that the 'audience' for digital media tends simultaneously to be situated everywhere and nowhere. Thus, building on the above arguments, and the notion of the 'pull focus', audiences (like 'children') are hyper-*visible* whilst the actual practices, feelings and meanings generated by audience-subjects are effaced. As much as a methodological gap between attempts – ironically, like that in the next part of the chapter – to visualise digital social networks and more ethnographic approaches (e.g. Paakkari et al., 2019), there is here a contradiction, if not a crisis of visibility:

> this enhanced visibility obscures more than it reveals. In data visualizations of audiences' (or users') activities, much of importance is stripped away. Away with the socio-cultural, displaced by individual "behaviors". Away with context, meaning, interpretation, for it is the hidden patterns beneath awareness that matter. Away with audiences' motivations, commitments, and concerns – for if data reveal what people "really" do on and through digital media, why talk to them anymore? [. . .] At present, the distance between real and data selves is often great – witness the academic uses of big data that fail even to distinguish men or women, adults or children.
>
> (Livingstone, 2019: 176–177)

Critically, Livingstone emphasises that just as impressive graphic visualisations of networked audiences or social media sentiment analyses may not tell us all we need to know about digital cultures, nor do ethnographic approaches (a point reinforced in a moment). Hence, perhaps twisting this argument a little, I argue through the examples in this chapter that there is a need to analyse how child-hoods *are* visualised and circulate in and through social media in a way that may not distinguish the ages of users but nevertheless in which childhoods/childhood concerns are clearly articulated – a task with which most research on digital media and young people has generally not engaged. Additionally, there is a need to inte-grate or, at least, juxtapose such analyses alongside ethnographic and other ways of making visible children and childhoods.

A rather different way of thinking about visualisation is to be found in a glut of recent research on the 'datafication' of childhood. I will return to these litera-tures in more depth below, but note here that, in keeping with the idea of the 'pull focus', there is once again a sense of ambivalence or tension. There is in this work a tense oscillation between the benefits of data collection (such as increased lev-els of automation, machine-learning and end-user satisfaction) and surveillance (such as purported benefits for children's safety and parents'/carers' peace-of-mind), and ongoing problematisations of the intrusions into individuals' privacy that surveillance can bring, with concerns about the collection of data – especially about children – and their use by large commercial enterprises such as Google (e.g. Finn, 2016; Mascheroni, 2018; Bradbury, 2019). Although important, I will not in this chapter add to recent calls to critique the effects of platform technolo-gies. Indeed, with particular emphasis upon schools, there is plenty of evidence to indicate how these technologies serve both to reorient teachers' practices and constitute new forms of 'transparency' in which children themselves are marked as 'progressing well', or 'normal' or giving 'cause for concern', thereby opening out or foreclosing future possibilities for an individual child (e.g. Kumar et al., 2019; Willson, 2018; Swist et al., 2019). Rather, whilst I do want to remain mind-ful of these concerns, I want – as a geographer – to tackle what are at present rather unresolved debates about the spatialities and scales of datafication: both in terms of claims that "[d]atafication leads to new spatio-temporal entanglements and transforms translocal relationships" and boundaries and that "classrooms [and other institutional spaces] are transformed from a physical location with teacher grade lists to a – to some extent – transparent and distributed datascape" (Jarke and Breiter, 2019: 3).

This latter claim leads into a final challenge for thinking about how digital media and childhoods are entangled: the status of *matter*, as and beyond the ways it has been articulated thus far in this book. Once again, this is, in part, a question of visibility. A key consideration in this light is the embeddedness of digital tech-nologies in many people's everyday lives, and their embeddedness in everyday objects – via smart toys and other constituents of the Internet of Things (IoT). And once again, there is ambivalence and tension: as digital technologies become ever-smaller and implanted within ever-more objects, the visibility of those tech-nologies (and our knowing interface with them) diminishes (Plowman, 2019); meanwhile, just as the focus seems to turn to the 'things' in the IoT, the impor-tance of the object-ness of objects – like toys – also seems to recede, as what matters more is the network that enables flows of data rather than, necessarily, the thing-ness of individual things. Of course, ANT has always emphasised networks more than objects, but there is a sense that things such as smart toys barely make it as objects (or actants), or have their objectness transformed to the status of mere conduit, incidental to processes of datafication. As I indicated earlier, a corollary of this observation is, as Livingstone (2019) points out, that, in attempts to visualise the ceaseless flow of data, many academic analyses of digital technologies and cultures have tended to efface the audience and the material cultures of which they are part. In other words, at the same time as objects recede in attempts to visualise data flows or critique the ethics of datafication, the embodied and material experiences of humans are also rendered less visible.

We are, therefore, caught up in a complex series of apparent contradictions – not least in terms of the status of matter (and, especially, objects) in digital cultures, which might require a bit of a rethink of both new materialist and object-oriented conceptualisations of matter. The same goes for humans (and, especially, children). One response might be to return to notions of hybridity and the cyborg – to recognise that some aspects of human life are ineluctably technological. Another is, as Livingstone argues, to place audience-centred research alongside other attempts to make digital technologies visible.

A third possibility is to focus on the *interfaces* that emerge through digital cultures (Rose, 2016; Ash et al., 2018). In this approach, objects remain 'visible' to a degree – indeed, object-orientation (as per OOO) can help in an understanding of how digital-cultural interfaces work, albeit in communications that are always partial. Thus, adopting OOO as part of a post-phenomenological perspective, Ash et al. (2018: 169) write:

> post-phenomenological approach to interfaces focuses on understanding an iOS app like Facebook not as a single thing made up of multiple parts, such as a map, a comment box and a series of sound effects, but as a set of multiple objects, where each object such as the map or comment box or sound effect exceed their relationship to the interface as a whole. From this position, interfaces are sets of objects that communicate partially with one another in order to produce the effect of being a single or coherent thing for the user.

As Ash et al. (2018) put it, analysing the interfaces of digital media – which may or may not visibly include humans – is not like analysing a painting, with its series of fixed parts. Rather, the units (perhaps, objects) in interfaces encounter each other in different, partial ways. Using the example of How-Old.net – an app that can age a person from a photo – they suggest that particular qualities of a photo are activated in its interaction with the app's algorithm that might not be when, say, a human attempts to age that photo or even another algorithm (such as facial recognition software used by law enforcement). In this way, as Ash et al. (2018: 172) argue, their approach is dedicated not simply to making-visible the hitherto invisible (and thus often purportedly insidious) machinations of digital media but to attempting to understand encounters at interfaces involving digital media, and much else besides. I develop and extend this approach to interfaces through a consideration of objects advertised on the buying/selling platform of an online selling site – both in this chapter and in Chapter 6. I also note that the different local manifestations between smoke and aeroplanes might – especially in their differential mediation – also be understood as interfaces.

This brief foray through different literatures and theorisations of digital media – and the (in)visibilities with which they have been concerned – has been necessary to frame the subsequent analyses and arguments in this chapter. This is because, on the one hand, the rest of this chapter will – drawing on some of the approaches detailed towards the end of this section – seek to move beyond common approaches to *children* and digital media, and especially work on risks and identity-formation. Thus – although beginning with the kinds of analyses of Twitter and an online selling site of which Livingstone (2019), at least, has been critical – I juxtapose a series of perhaps slightly oblique ways in which it might be possible to visualise children and digital media. On the other hand, the above literatures witness a series of ongoing but, in my mind, generative ambivalences and tensions – centred on the complex relationships between, and imbrications of, digital media and matter (and especially objects). Drawing on some of the latter points raised in this section, I redeploy Bryant's ontology of machines and media – alongside other theoretical tools – to offer further ways to work through these tensions and to open out a far broader repertoire of approaches for analysing *interfaces* between and beyond children and digital media.

Childhoods, circulating I: Twitter, an online selling platform and dolls (again)

It seems an obvious point – at least for anyone familiar with the New Social Studies of Childhood – that discourses about children and childhood *circulate*. The social construction of childhood is, perhaps at its heart, a process of creating, visualising and disseminating representations of childhood. Indeed, boiled down to their fundaments, the examples in this section form an extension of these processes. In making an argument about childhoods, circulating, I am not claiming that there is anything fundamentally *new* going on. Like Lee and Motzkau (2011), however, I *am* claiming that there is something about contemporary technologies – in this case, digital media – which means we are witnessing not only new biosocial or biopolitical formations but *new and intensifying modalities* through which childhoods circulate (also Swist et al., 2019). I am also claiming that – surprisingly, given the range of work on children and digital media, and on the social construction of childhoods – there have been few, if any, analyses that work *with* those new modalities in order to articulate *how* childhoods circulate through, for instance, social media.

This section of this chapter offers two sets of responses to this lacuna. Both of these responses draw on and extend the analyses of #AngerForAleppo, presented at the beginning of this chapter; both are concerned with the circulation of some 'stuff' – talk and material objects – about childhoods; and both are orientated towards a sense in which, in the process, childhoods themselves circulate. The first focuses on Twitter; the second on an online selling platform.

As part of the Plastic Childhoods project, we had the opportunity to harvest and analyse, via an API, a range of non-plastic-related hashtags on Twitter that nevertheless related to children and childhood. We chose hashtags that were either trending during the main period of data collection (November 2018–April 2019) or that were pertinent to current affairs or events during that period. One such set of hashtags centred around the #climatestrike events. I look at these events in more detail in Chapter 6, although I make a couple of observations in this chapter that relate to the examples presented there. Here, I first focus on a topic that has been of increasing interest in both policy and academic circles, not least in relation to digital media (Goodyear and Armour, 2018): children's mental health. Figures 5.1 and 5.2 show visualisations for one of the main hashtags – #childrens mentalhealthweek – between 15 February and 7 March 2019. Children's mental health week itself took place (in the United Kingdom) during the middle of February and involved a range of activities, events and resources for schools, parents and young people themselves.

Figures 5.1 and 5.2 present striking graphics and, notably, contrast significantly with Figures 6.1–6.3 (tweet activity around #climatestrike, discussed in detail in Chapter 6). Figure 5.1 shows the hashtag network analysis for the period specified above. Each grey dot represents an individual user and the size of the dot the level of their activity (tweets and retweets). The grey arrows represent social

#childrensmentalhealthweek

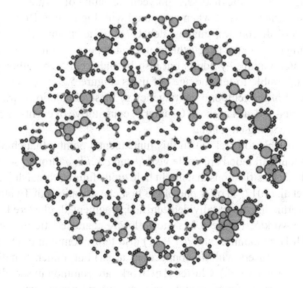

Figure 5.1 Hashtag network analysis for #childrensmentalhealthweek

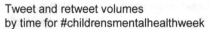

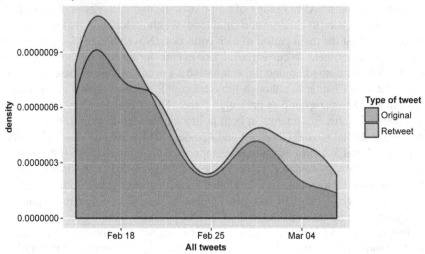

Figure 5.2 Tweet density analysis for #childrensmentalhealthweek

networks (retweets or replies) and are therefore connected to other users; the larger the number of arrows, the greater the density of activity around any particular user(s) (contrast Figure 5.1 with Figure 6.1). Indeed, the most noteworthy feature of Figure 5.1 is the disparate, polycentric nature of Twitter activity around #childrensmentalhealthweek when compared with Figure 6.2. Both figures offer visualisations of discourses, images, debates and viewpoints *about childhoods*. Upon scrolling through the tweets themselves, it appears that the constitutive content was created mainly by adults but also by older children, although in some cases it is impossible to tell the ages of users. Figure 5.2 (like Figure 6.3) also affords a sense of the temporal contingency of these circulations, which may be fairly short-lived and, as I argue in Chapter 6, constitutive of events or even 'infra-generations' involving childhood.

It is then possible to break down this fundamental point – that childhoods and discourses about childhood are made *visible* and *circulate* through social media – into conceptual classifications that might help guide future scholarship in this area. Himelboim et al. (2017: 1) offer one of the first classifications of Twitter networks based on information flow: "divided, unified, fragmented, clustered, [and] hub-and-spoke networks". Figure 5.1 is most closely characteristic of *clustered* networks, which Himelboim et al. (2017: 11) also term 'community clusters': there are few 'isolates' (users with no connections at all) but, rather, 'small groups of interconnected users' (ibid.). Clustered networks are common in social media networks surrounding conferences and social or cultural programmes – #childrens mentalhealthweek would therefore appear to be exemplary of this kind of network.

Clusters are indicative of a larger number of small, focused discussions, often representing a diversity of opinions.

Meanwhile, Figure 6.2 is very different, and Himelboim et al.'s taxonomy aligns most closely with 'hub-and-spoke' networks. In this kind of network, Himelboim et al. theorise that there is a kind of 'broadcast' element – akin to fairly traditional modes of broadcasting from the second half of the twentieth century – in which large numbers of 'audience' members either subscribe to or endorse the representations of a small number of individuals. In Figure 6.2, for instance, we can immediately surmise that part of the reasoning for this difference is that the #climatestrikes have been driven by influential individuals (such as Greta Thunberg). Interestingly, whilst celebrity Instagram users, Vloggers and others do map onto this broadcast model of the hub-and-spoke network, the #climatestrikes are slightly different, on two fronts: Greta Thunberg and other key users are not positioned as 'broadcasters' as such – rather as key conduits or points of articulation for a social movement that takes place as much offline and online involving collaborators rather than 'audiences' (compare Livingstone, 2019); and if we look into the minutiae (of all tweets, in Figure 6.1, rather than all tweets with $n = 10$ retweets/replies or more, in Figure 6.2), we see a large number of what we could term 'outliers'. These outliers are in part indicative of 'clustered' networks but – because many users are isolated, with no connections at all, are also symptomatic of 'fragmented' networks.

Whilst Himelboim et al.'s (2017) conceptualisation of social media networks is a useful starting point for characterising *how* childhoods circulate, clearly there is more to be done. Indeed, although offering a more precise sense of the size and composition of socially mediated 'stuff' about childhoods, there is arguably (and perhaps ironically) not that much specific to *childhood* – whether globally or locally – in the previous paragraphs. One way to offer more conceptual and empirical specificity would be to delve into the very details of individual Tweets – something I do in Chapter 6 in discussing the #climatestrikes in more depth. Another would be to engage in sentiment analyses of the whole set of tweets associated with a particular hashtag during a certain period, in order to gain a sense of whether activity around a particular issue is broadly positive or negative. On this front, of the sample of 20 hashtags we analysed over the same period in early 2019 for the Plastic Childhoods project, only two (#plastictoys and #plasticoceans) were negative overall, with the remainder being positive. The net score (subtracting negative from positive tweets) was, however, far higher for #childrensmentalhealthweek than #climatestrike, indicating something of the backlash associated with the latter amongst (mainly) adults critical of this and likely *any* form of youth action, and/or of children taking time out of school, and/or of the very idea of climate change. However interesting – and however much they *might* offer further insights into the broad feelings associated with children – sentiment analyses remain fairly rudimentary, and most analysts agree that they are a fairly poor proxy for users' emotional intent and/or responses when using social media (e.g. Bravo-Marquez et al., 2014).

A third way to reflect on #childrensmentalhealthweek and #climatestrikes is to compare the modalities of circulation witnessed in this chapter with a small line of work – concentrated ostensibly around a special issue – that has specifically examined the notion of 'circulation' in childhood studies scholarship. Amongst – and somewhat problematically not explicitly cognisant of – significant attention to children's *mobilities*, Stryker and Yngvesson (2013) offer a fairly brief exposition of what a theory of circulation might offer to childhood studies. Critical in their view is the sense in which circulation is not simply bounded – as it might be when one considers the circulation of blood around the body. For them, a conceptualisation of circulation that embraces both fluidity and fixity, rather than fixing children in particular spaces, offers the grounding for a 'phenomenology of circulation'. Here, "[c]irculation is also used as a metaphor for understanding children's experiences of movement and space, both their own circulation and the circulation of ideas, narratives, images, and symbols associated with childhood" (Stryker and Yngvesson, 2013: 298). Their argument, and the papers that follow in the special issue, tend towards a phenomenology of circulation (understanding children's experiences of circulation and how these contradict fixed notions of childhood) and the deployment of circulation as a metaphor (for instance, in the 'circulation' of memories around an institution). They have less to say, however, about the circulation of ideas, narratives (etc.) that exceed such metaphorical or semantic delineations of particular phenomenological experiences, let alone the increasingly important role of digital media *in* their circulation. In other words, I agree wholeheartedly that we require a more nuanced theory of circulation. Yet, in order to broach the new and intensifying modalities *of* digital media, we also require – in Ash et al.'s (2018) terms – post-phenomenological analyses of childhoods, circulating.

In the rest of this section, then, I want to turn to the notion of the *interface* – albeit in a slightly different way than either Rose (2016) or Ash et al. (2018) use it – in order to begin to grapple with some of the questions around (im)materiality raised earlier in the chapter. Alongside the tweets, during the Plastic Childhoods project, we analysed a purposive sample of objects associated with childhood appearing on a well-known, globally available online selling platform. In fact, we were not – initially – interested in *specific* individual objects but in *categories* and *sub-categories* of objects, which we used as search terms when running the API that harvested our data from the online selling site. The categories of object were 'cars', 'dolls', 'baby', 'lunchbox' and 'miniature'. We then used a painstaking iterative process (outlined in Chapter 1) of analysing the pictorial and textual comment for individual items for sale alongside a more refined meta-level content analysis (via the API) that enabled us to discern certain sub-categories. Whilst far from perfect, these categories and sub-categories give an insight into the kinds of *stuff circulating about childhoods* that would, I suggest, be very difficult to glean through other media. Specifically, it would be difficult to find other ways to witness the sheer *scale* of stuff circulating about childhoods as that evident in Table 5.1, which refers to over six million individual items for sale (from our sample of just five categories of objects).

Before going on, I should note that our positioning as a research team – as white, middle-aged men, both fathers of fairly young children and both located in

Table 5.1 Categories and sub-categories of childhood-related objects returned in from an API scrape of an online selling platform, March–April 2019

Category	Sub-category (and refined API search term)	Total results	Items listed as new	Proportion as new
Cars	Toy cars	33,825	25,310	74.8%
	Vintage toy car	3,812		
	Job lot cars	937		
Dolls	Doll	871,897	522,449	59.9%
	Reborn doll	31,020	25,746	83.0%
	Job lot dolls	724		
	Vintage doll	11,273		
Baby	Baby	4,599,943	4,103,869	89.2%
	Baby bottle	17,265	15,539	90.0%
	Baby feeding	427,315	347,458	81.3%
	Baby clothes	382,094	348,210	91.1%
	Baby pram	50,123	45,519	90.8%
Lunchbox	Lunchbox	118,568	108,282	91.3%
	Children's lunchbox	10,176	9,869	97.0%
	Plastic lunchbox	14,924	14,245	95.5%
	Miniature toys	33,523	30,439	90.8%
	Miniature furniture	71,084	65,641	92.3%
Miniature	Miniature furniture vintage	8,593		
	Miniature weapons	812	576	70.9%
	Miniature cookware	1,530	1,454	95.0%
TOTAL		6,689,438	5,664,606	86.2%

the United Kingdom – will clearly have influenced our initial selection of object categories and the sub-categories that followed. I therefore make no claims as to the universality of these objects, although I hope that they resonate with at least some readers.

Notwithstanding the above caveat, it is possible to make some broader conceptual claims about the objects in Table 5.1. Foremost, they offer part of what I hope is a gathering sense of the ways in which childhoods circulate in modes that extend beyond the phenomenological (i.e. wherein child*ren* are decentred somewhat but nonetheless remain in [pull] focus). Specifically, through different (social) media, it is also possible to make *visible* particular kinds of circulations of *stuff* about childhoods – both in the form of topics (hashtags) and their broad sentiments, and in the form of material stuff like – literally – *millions* of dolls and bay items. Critically, these forms of visualisation are not somehow 'better than' established studies of children's material or popular cultures; rather, they simply offer a different 'take' on all that stuff – one that works *with* the media that supports its circulation as a way to *visualise* its circulation.

As with Figures 5.1 and 5.2, the conceptual and analytical possibilities that emerge from data such as those we collected from an online selling site are – especially in combination with other methods (Livingstone, 2019) – virtually limitless and virtually un-explored by childhood studies scholars. I want to focus on just one set of possibilities, concentrating on *interfaces* and inspired by Ash et al.'s

(2018) analysis. When considering the particular online selling platform that we worked with, it is important to remember that individual objects are encountered both at particular moments in their own (individual) circulation and as part of the circulation of *masses* of similarly categorised stuff. In other words, the layout of platforms like this is one that deliberately enables interfaces with objects in the singular *and* the (massified) plural.

Firstly, then, it is helpful to ask how objects are *encountered* through the site's interfaces. When logging on (in August 2019), the *tone* of the site is fairly straightforward, because the different *units* – the search function and the scrolling, coloured banners offering discounts – afford two sets of simple ways to interface with the site (following Ash et al., 2018). Things become a little more complex – if not bewildering – from there, however. After entering the term 'doll' into the search function – and yielding 939,252 results – images of dolls are returned in no particular order, beyond sponsored adverts. However, the site's own category unit – indicating the generic status of an object as 'pre-owned', 'new' or whatever – affords the user some opportunities to categorise and therefore refine their search. Complicating the interface with a human user are the ways in which that interface *vibrates*, as Ash et al. (2018) have it, in ways that may or may not condition your choice of doll – most specifically in the form of a 'time left' countdown timer that indicates when the auction for that particular item closes. Upon finally finding your ideal doll after (potentially) many hours of searching, you click through to the item itself, to be faced with some very particular ways of interfacing with the object in question (examples in Figures 5.3 and 5.4). None of these can, to paraphrase Harman (2011a), exhaust the doll in its very being, because they offer strictly limited ways to interact with the qualities of an object.

These modes of interfacing include the restricted capacities of the platform itself (its still fairly rudimentary, almost purposefully rough-and-ready interface), the qualities of the photographs, if they are present (and, in turn, thinking of Figure 5.3, the quality of the photographic equipment used, the ambient lighting and its play across surface textures, the background, the skill of the photographer) and the qualities of the textual descriptions (and, in turn, thinking of Figure 5.4, the

Bidding has ended on this item.

Zapf - Baby born - black doll in played with condition and was much loved! See original listing

Condition: **Used**
Ended: 10 Feb, 2019 20:19:12 GMT
Winning bid: **£4.70** [6 bids]
Postage: **£3.25** Economy Delivery
Item location: Lincoln, United Kingdom
Seller: Seller's other items

Sell one like this

Figure 5.3 Screenshot from an online selling platform showing a doll returned from our API search. Caption Reads 'ZAPF – Baby born – black doll in played with condition was much loved!'

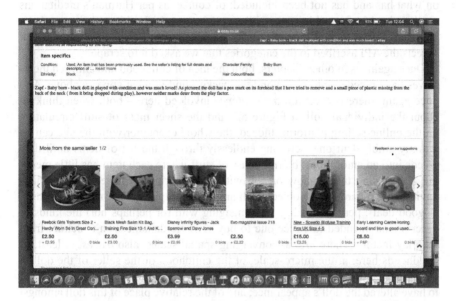

Figure 5.4 Screenshot from 'item description' in an online selling platform page for the same doll as in Figure 5.3 containing further details. Alongside specifics about the doll's ethnicity, brand and hair colour, the free-text description reads 'Zapf – Baby born – black doll in played with condition was much loved! As pictured the doll has a pen mark on its forehead that I have tried to remove and a small piece of plastic missing from the back of the neck (from it being dropped during play), however neither marks deter from the play factor'.

relationship between the text and the object, the properties of the object that are chosen in an often short description and the skill of the writer). Ultimately, all of these *tones* (Ash et al., 2018) will – in tandem with your own preferences, feelings, budget, cultural norms, the identity of the intended end-user – affect whether you click the big blue button that reads 'Submit bid'.

In turn, there are other ways of encountering this doll and the millions of other plastic items for sale on this platform wherein the human recedes somewhat and where partial interfaces involve object-to-object encounters. These are, for instance, evident in the ways *we* encounter the doll on the site: in the pen marks on its forehead that, despite the best efforts of the seller, are, given the properties of the pen and the plastic from which the doll is forged, indelible; or in the small piece of plastic missing as a result of coming into contact with an unspecified (doubtless harder or sharper) surface when the doll was dropped. These are also evident, for instance, in the way in which our automated API search encountered objects. For, as in Ash et al.'s (2018) example of How-Old.net, the API interacts with objects (and, specifically, the units on each selling page that make-up that object's digital alter-ego) – looking both for text corresponding with our search terms but then categorising objects according to the properties that the site has allocated to it (such as 'new'). A simple corollary of these encounters is to reflect on what has and has not been included: of course, as per Harman's meditations on the 'depth' of an object, we will never know which objects have *not* been included – because they were not returned in our search – but we do know which objects the API ascribed to the categories that we asked it to scrape.

Once again, as in other chapters, both the idea of childhood and children themselves have receded into or beyond the background in what appears above. And once again, there is a certain *ambivalence* involved here – both when thinking about the individual doll in Figure 5.3 and the sheer mass of stuff circulating on the online selling platform. Indeed, the whole encounter with the site can be a pretty ambivalent one: scrolling endlessly through the list of items until they recede into an amorphous, blurred mass of stuff where each item has little meaning and closing the window or app without bothering to place a bid; perhaps having little regard for the object itself and its biography in the interest of getting a 'good deal'; perhaps reading and engaging with (or perhaps not) the kinds of descriptions intended to entice one to buy – where damage won't 'detract from the play factor'. There *are*, at converging spatial scales, also *traces*, at least, of childhoods here: at the micro-scale, of the childhood of the seller of the doll in Figure 5.3, and at least two moments of play (or inattention, or both) that appear to have altered the doll's appearance; and of the relative place of this doll amongst a range of cultural and commercial norms surrounding childhoods (where black dolls may be less common in some contexts, or where expectations around the gender or clothing of a doll reinforce gender norms in others). On the former, I say more about objects' biographies and their role in what I term 'infra-generational' accounts, after childhood, in Chapter 6. On the latter, I reflect further on questions of race in Chapter 7 but also note that – beyond the bounds of this book – there is significant scope to deploy the iterative approaches to analysing an online selling site and other similar digital media developed here in more critical accounts that interrogate the ways in which cultural norms, expectations and forms of marginalisation are expressed and experienced through digital cultures and their interfaces.

The inclusion of a doll as an example in this chapter has not only served as a means to think again, and to think differently about a common (and often contested) childhood object; it also offers an important point of connection back to the detailed conceptual discussions of matter introduced in the previous chapter. Bryant's notions of machines and, especially media, offer a powerful frame for juxtaposing the doll found in the apartment in Luz and the one found on the online selling platform. In both cases – and aligning with Ash et al.'s (2018) language – the dolls are units that are both composed of other units (or entities) and that in turn compose still others (Bryant, 2014). In both cases, in these nested sets of relations, the dolls operate as, and as part of, machines: their interfaces differ wildly (and I have offered significant details of these interfaces in my analyses of the two dolls); yet they both *operate* dynamically, in the production of amorphous inputs and outputs. But the dolls are still vastly different. They impinge and insist power upon humans in different ways; they are machines of a broadly common lineage that have been re-programmed to operate upon humans, and our emotions, in different ways (all the while the association with childhood is a central element of those operations). In principle, this is a point about locally manifested interfaces, but I think it is more powerfully articulated as a point about *mediation* (and, again, *visualisation* and *circulation*), where media are considered as 'medianatures' (Parikka, 2013) or 'posthuman media ecologies' (Bryant, 2014). Media – or intermediaries – extend beyond popular or digital media but, at the same time, may include them. And when they do, the interfaces that are afforded are very different than in those cases (as in Luz), where digital or social media are less evident. In this way, media – like dolls (conceived as machines with inputs and outputs) – are productive rather than representational, both exerting and channelling local manifestations of power, although in the case of the doll found on the online selling platform it is less clear 'where' that 'local' is.

More straightforwardly, Morton's (2013) notion of hyperobjects might help account for the *massification* of stuff *circulating*, about childhood, in a way that builds on other attempts to *visualise* and narrate the elusive, blurrily bounded, stuffness of stuff (e.g. Horton and Kraftl, 2018). Dolls – either as a global phenomenon themselves or as a species of the genus 'plastic' – could be seen as a hyperobject. And, although like other objects, they cannot be perceived or exhausted in their totality, the very different encounters in Luz and on an online selling site afford glimpses of the 'phasing' of dolls' objecthood into and out of human experience (Morton, 2013). Moreover, large-scale API analyses offer a different perspective onto the very hyper-ness of the hyperobject doll/plastic: not a clearer or more accurate view but certainly a making-visible of the sheer *scale* of the object in ways that complement encounters with individual dolls and that make-visible that sense of sheer mass one gets when logging onto online selling sites. Both offer different senses of the 'pull focus'; juxtaposed, the stories from Luz and the online selling platform represent a simultaneous zooming in and pulling away, yet both distort what we already know about dolls and their 'meaning'. And both offer different perspectives of the ineluctable entanglement of hyperobjects with and *as* 'us' humans – not least in the poignancy of the doll left behind in Luz, and its unknown owner, or the brief glimpse of a long and caring history of ownership and play found in the two-line textual description on an online selling site.

In this section, I have sought to offer steps along the way to theorising childhoods, materialities and digital media *together*. Through the tropes of circulation, network, interface and visibility, I have considered a suite of ways in which we might work *with* the speed, intensity and sheer scale of digital (especially social) media in order to continue thinking and doing, *after childhood*. Alongside a range of other conceptual perspectives, notions of media, machines and hyperobjects could prove to be helpful for theorising the workings, effects and affects of childhoods, circulating.

Childhoods, circulating II: artificial intelligence, the IoT and infrastructures

The examples in the previous section demonstrated not only how childhoods circulate but, more specifically, how those circulations are composed of both *bits of* childhoods and, simultaneously, *bits associated with* childhoods. Dolls are a good example: as *bits of* childhoods, they comprise extensions of childhood experience and, indeed, as Aitken and Herman (1997) showed, are one of a number of 'transitional' objects that help a child make sense of themselves and the world. And as *bits associated with* childhoods they are – in the collective as well as the particular – some of the most enduring and 'obvious' objects. Yet, to repeat and extend Livingstone's (2019) point, neither these material objects nor digital traces correspond entirely with (in her terms) 'real' selves or subjects. Connecting Livingstone's argument with OOO and speculative-realist philosophies, this is – in a flat ontology – because just as with any other object or machine, human bodies can themselves not be exhausted, actualised or visualised in their depth or extent (as Deleuze and Guattari [1988] also famously argued through their concept of the Body without Organs).

Continuing these lines of thought, the purpose of this section is twofold: first, through juxtaposing a range of fairly brief examples taken from contemporary media and academic coverage about AI and the IoT, to tackle one of the key debates about contemporary childhoods: datafication. Datafication is defined as the increasing accounting for, and reduction of, information about children (such as their learning) to databases, software and attendant surveillance practices, and the increasing reliance upon such data for dealing with (usually teaching) children (Finn, 2016). The second is to indicate how analyses of datafication could be supplemented with and connected to theorisations of matter and social difference. Specifically, I want to develop predominantly Foucault-inspired analyses of power, visibility and surveillance that theorise "how data can reproduce the child as a 'data double', stripped of all complexity, and how this dividual can be subject to visibility rather than the person" (Bradbury, 2019: 10). In part, as Gulson and Webb (2017: 9) argue, this means critiquing the

> transformation of biology to a data-driven and computing-based field has seen the entangling of computing power, data and new kinds of technical infrastructures with a focus on statistics and sequencing. These entanglements are part of 'bioinformatics', a term used interchangeably with 'computational biology' and 'systems biology'.

Yet, as I have already argued – through the plural notion of 'interface' – these infrastructures are not merely technical (in the digital sense) and nor are they confined to questions of the biological and to the geological (Gallagher, 2019; see Chapter 4) but to, as I will show in this chapter and others, the banality of everyday childhood objects like toys.

The rise of AI products – especially in education – has been seen as an intensification of efforts to 'personalise' learning in (particularly) neoliberal contexts (Cukurova et al., 2019; also Pykett, 2009; Perrotta and Williamson, 2018). Boiled down, AI can provide real-time support and recommendations for learning content, speed and method, at the same time as – recursively – collecting data about children, both individually and collectively. It can also – as will become clear – offer different interfaces for learning although, as Thompson and Sellars (2018) argue, the status of these technologies as comprising an event that will 'rupture' or 'revolutionise' learning is rather less certain. As part of the Energy Beyond Technology project, we developed systematic reviews of academic literature, (social) media coverage and commercial websites dedicated to datafication, AI, IoT and algorithms. I focus here on the features of a subset of these technologies, in order to continue developing a conceptual framework around the *circulation of stuff about childhoods* that is attuned to both the digitality and materiality of these technologies.

As indicated above and in the growing literatures on datafication and AI, the education sector is seeing perhaps the most concentrated growth in digital technologies aimed either at children themselves or at generating data to support teachers in their work with children (Kumar et al., 2019; Gulson and Sellar, 2019). Many of these technologies are produced by commercial companies, prompting a range of fears and concerns about the ultimate uses of the data produced (see below) and "a redistribution of agency across socio-technical networks" that involves a range of 'new' actors (Williamson, 2017; Jarke and Breiter, 2019. A couple of examples will suffice.

Firstly, Kidaptive Adaptive Learning Platform (ALP) is a learning platform service for other educational companies to analyse the data produced by teachers and pupils using the platform in order to 'improve learner engagement and outcomes' (www.kidaptive.com/). It is a cloud-hosted Big Data platform which, Kidaptive argues, makes it scalable and secure. Using a bespoke machine-learning algorithm it can be used to create personalised learning paths that are communicated to children and parents; identify learning-relevant behaviours like guessing, focusing and skipping; set appropriate challenges; track skills learnt by learner during game-play as evidence for educators; and help test preparation through identifying gaps in learning and through personalised learning. The ALP works through three tiers: Analytics ("One-way event sending to ALP, which provides analytics based on event data [. . . such as] engagement and retention"); Insights ("metrics, ability insights and insights [. . . with ability to] predict test performance, estimate time on task, and more"); and Adaptivity ("support for real-time client-side adaptivity [. . . using] a Bayesian-IRT psychometric framework to determine the optimal next challenge") (www.kidaptive.com/). There are many other similar platforms, such as ALEKS (Assessment and LEarning in Knowledge Spaces). This is an AI assessment and learning system to determine quickly and accurately knowledge of students. ALEKS periodically assesses the student as they progress through a course, providing instruction and directed assessment to ensure learning is retained. It is available for maths, business, science and behavioural science subjects in America (www.aleks.com/about_aleks).

One of the most common responses to AI and the increasingly pervasive 'atmospheres of progress' (Finn, 2016) that ensue in schools has been to outline concerns not only with the rise of commercial companies as providers of platforms but the security and ultimate use of data circulating about children. Yet there are many other possibilities for explaining and conceptualising these kinds of digital technologies. On the one hand, there is a need to theorise the positioning of childhoods – and children – within an increasingly evident 'platform society' (Van Dijck et al., 2018). "The term refers to a society in which social and economic traffic is increasingly channelled by an (overwhelmingly corporate) global online platform ecosystem that is driven by platforms and fuelled by data" (Van Dijck et al., 2018: 4). A platform is a digital structure in which users can interact and where data about users can be collected and afforded financial value (e.g. Uber or ALEKS); a platform ecology refers to the system of platforms that predominate in any context – in countries like the United Kingdom and the United States, these include Amazon, Google, Uber, Facebook and so on.

Kumar et al. (2019) offer one of the first systematic conceptualisations of platforms that can help us understand the positioning of children within digital media such as Kidaptive and ALEKS. They argue that platform systems encourage teachers to become 'surveillant consumers' of digital technologies and orient them "to see student data as interchangeable with students" (Kumar et al., 2019: 1). Once again, then, platforms are dedicated to *visibility*; but, once again, rather than making-visible students' own experiences or voices, these systems move *beyond* voice and agency towards the constitution of datasets *about* individual children or groups (also Kraftl, 2013a). Those datasets can then take on their own lives, both figuratively and literally far from the children with whom they originated. The vast majority of such datasets remains obscured (*in*visible) but is then made visible in partial ways to both teachers and students, through the carefully designed *interfaces* of any particular platform. Thus, not only do data selves correspond poorly with 'real' selves (Livingstone, 2019) but children themselves only ever see partial and heavily policed versions *of* those data selves. As Finn (2016) and Kumar et al. (2019) argue, these forms of representative have significant effects and affects: some learners may be pleased with their data; others anxious or scared; others will ignore it. But now that data accompany children throughout their whole childhood, into adulthood, it is becoming increasingly possible that those data will govern the opportunities (and limitations) presented to any child as much as any other considerations (Bradbury, 2019).

Kumar et al.'s (2019) analysis is particularly helpful for the ways in which it offers additional tools for thinking and doing, after childhood. My reading of their work (and I stress that this is my interpretation) is that it offers a conceptual lexicon for thinking more about platforms in education, as follows. Data are *all-consuming*; they chase teachers, even as they sleep, pressing for a response. This affords, perhaps more than anything, a sense in which data are *out of the total control* of not only teachers but also the technology companies that create them. This is because the data are so complex, so massive (even for just one school) and because data are aggregated and reported back through machine-learning that can take place with little measure of human intervention. Meanwhile, where lack of information is now seen as a threat, platforms can – because of their interoperability across multiple devices and places – offer possibilities for 'eliminating *friction* and *flattening* students into data' (Kumar et al., 2019: 6, my emphases). This argument is, then, not only about the difference between real and digital selves but the *flattening* of students into 'one-dimensional units within a uniform inter-face' (ibid.: 7). Thus, the very vibrations of the interface (Ash et al., 2018) mean that, increasingly, teachers run the risk of confusing their knowledge of data with knowledge of their students. However, "[t[he platformization of the classroom does not cause teachers to monitor their students. Rather, it reduces the friction involved in monitoring and frames monitoring as an *attractive* method through which teachers can perform their duties" (ibid.: 7; my emphasis). In other words, neither teachers nor students are cultural dupes – and as I know from my experi-ence of working in many schools that use many different methods to collect and deploy data – neither do they *not* know their students, beyond their data. How-ever, the platforms, their interfaces and wider professional and societal expecta-tions and norms around the 'threat' of data-lack mean that they offer a plethora of attractions for teachers. Thus, platforms are all-consuming, out of (total) control of humans, may eliminate friction and flatten students, and may seduce and attract teachers (and others) through their interfaces.

Certainly, of the above concepts, the central idea that digital technologies and data are *out of control* is one that has gained widespread attention. Recent cases involving networked objects – most commonly referred to as the Internet of Things (IoT) – have raised many critiques and fears about the rights of consum-ers and the relative control that we all have over our data (Livingstone and Third, 2017). Some of the most high-profile examples have involved products for chil-dren. At the time of writing, the most prominent centred on Amazon's Alexa Echo Dot, which had been collecting voice recordings from hundreds of thousands of children, despite a lack of clarity about consent for doing so (BBC, 2019a). Whilst the recordings can be deleted by parents, it is likely that many will not have been – not least because of the time involved and because Amazon uses the recordings to personalise its services.

Although in the case of the Echo Dot there have not been any direct concerns over security, other examples highlight that the utopia of 'frictionless' data (Kumar et al., 2019) is one that requires considerable scrutiny, and especially where the technological elements of IoT are becoming increasingly invisible (Plowman, 2019). For instance, of many similar cases, a company specialising in internet-connected teddy bears (CloudPets) leaked over 800,000 user credentials in 2017 – exposing the problems of hacking and smart toys (Franceschi-Bicchierai, 2017). CloudPets are smart toys which enable families to record and leave voice mes-sages on a teddy bear, uploaded via bluetooth. Spiral Toys company left the data-base of credentials without firewall or password protection, exposing data which was then hacked and held to ransom (ibid.). Although the voice messages were on the exposed database, other researchers have found the messages were also stored without authentication (ibid.) and therefore easily accessible for hackers. These toys have since been banned in several countries, including Germany, and removed from platforms such as Amazon and other prominent online selling plat-forms (BBC, 2018b).

Thus, a corollary of the concepts of *visibility*, *control* and *friction* – and sit-ting in the relation between them – is the ever-present threat of *exposure* and the potential *trauma* that that exposure might cause (Chapter 9). I am not sufficiently well versed in digital security to comment further; rather, and thinking again of modes of thinking and doing, after childhood, I would note again that all of these concepts are components of the broader *massification* of data and hence of the ways in which childhoods circulate. Arguably, the voice recordings held by Amazon are – even if they are left unmined – somewhat akin to the Mass Observation Archive, which collected and collated writings and other media by 'ordinary' British citizens between the late 1930s and the early 1950s (www.massobs.org.uk/) and again from 1981 onwards (including an everyday childhoods collection from 2013 to 2014, held at www.massobs.org.uk/about/news/192-the-everyday-childhoods-collection-is-now-on-figshare). However, there are also obvious differences, not least in terms of the levels of consent obtained for the latter and the huge size of the former, which is already many orders of magnitude larger in terms of the sheer number of children involved and the details held about them. But what is also significant is the fact that although human operatives are listening to a very small proportion of Amazon's recordings, most are – like other data collected by AI – simply either deposited in vast digital reserves or analysed through automated algorithms, enabling ever-more sophisticated forms of machine-learning, ostensibly without humans.

These databases are, then, empirical examples par excellence of one version of the idea of thinking and doing, 'after childhood'. Not only are children *still a concern* (as data-producers, potential victims and paying consumers), but children are also, and *simultaneously*, decentred as their data are literally and metaphorically displaced from their lives, from their view and from their control (indeed, perhaps the control of *any* humans). Furthermore, a little like the toys discussed in Chapter 6 – although in some ways not all like them – these recordings could exceed children's childhoods and, indeed, their own lifetimes and could be taken not only as a proxy for an entire generation of childhoods but constitute in themselves 'generations' of a sort (generations of childhood-data), with temporalities that interweave and interface with human generations but also exceed them. One question is, then – amongst many others – whether the massification of these recordings and the many other kinds of data produced through AI and IoT constitute hyperobjects; and/or, indeed, whether, as a result, it might be possible and useful to think of *childhoods, circulating* as hyperobjects – particularly, but not only, in their datafied form.

Looping back to the feminist, queer and critical race theories I introduced in Chapter 2, and with an eye to forms of physical, behavioural and emotional difference that I discuss in Chapter 7, a final set of conceptual considerations for AI and IoT is ethical and political and focused around questions of *inclusion*. In light of concerns about *visibility, control, friction* and *exposure*, several platforms and technologies now purport to offer users – and specifically children – opportunities to wrest back some measure of control, particularly by learning more about the workings of these media. Teens in AI is an organisation launched at the AI for Good Global Summit at the UN in May 2018 to support young people aged 12–18 to explore AI, machine-learning and data science (https://teensinai.

com/#about_us). The organisation hosts 'bootcamps', 'hackathons', workshops and talks and other events for young people to learn how to work with AI. Summer programmes take place in London but are very expensive (a ten-day course costs £2,150 and you need to bring a computer), and while fully funded places exist, the programme remains inaccessible to many young people.

The exclusivity and inaccessibility in terms of young people being able to access codes is increasingly recognised as something impacting the algorithms and AI being developed (because particular, human-inputted codes and values underlie at least the initial development of machine-learning). As the technology is created and designed by a homogeneous group of people, the resulting technologies are biased towards their (likely) homogeneous needs, propping up serious concerns that AI and other digital technologies could reinforce economic or social divides and, in particular, the opportunities opened out to children and young people during their lifetimes (Gulson and Webb, 2017; Bradbury, 2019). In response, AI4All (http://ai-4-all.org/) is an organisation which aims to provide mentorship and education so that underrepresented young people in North America can learn AI (ibid.). Underrepresented groups (such as girls, low-income students and 'youth of color' [Johnson, 2019]) are mentored through training programmes hosted at universities across the United States. The success or otherwise of these schemes has yet to be determined.

In a very different set of circumstances, there have been moves to design robots to support children with autism spectrum disorders to better interact and learn social skills from their teachers and therapists. The University of Luxembourg's QTrobot is used within therapy sessions to decrease discomfort for those with autism and increase a willingness to interact (Waltz, 2018). The robot is described as a 'cute tech-based intermediary' (Waltz, 2018), designed to teach those with autism spectrum disorders to interact more comfortably. A video released by the lab working on the robots shows a boy being instructed by the robot to place objects in certain arrangements and congratulating the boy when he completes the task; the boy's gaze is focused on the robot. During the research, the designers found that children directed their gaze towards the robot twice as long as towards the human; behaviours which commonly signify distress or anxiety occurred three times less with the robot than with sessions with the human.

Once again questions of *interface* come to the fore here. In this case, however, the interface between children with autism spectrum and the robot raises two distinct yet potentially troubling issues that require further consideration. First, in terms of the promise of 'inclusion', the concern here is – as with any of the AI technologies cited above – inclusion in *what*? Do the robots and other technologies simply promise the presence – and *visibility* – of an increased range of consumers, who as the AI technologies become ever-better-attuned to the nuances of diverse human experiences can be incorporated into logics of commercialisation and datafication (Thompson and Sellar, 2018)? Recursively, a range of questions could be raised about the intent lying behind efforts to use technologies that might attempt to normalise, channel or control neuro-atypical humans into what may be actually fairly conventional educational and societal norms. Of course, the ability of the robot to make a child feel comfortable or communicate could be a good thing; yet only where 'comfort' and forms of emotional literacy are not taken as a proxy for the subjection of children with behavioural, emotional or social differences to power-laden educational practices that may – alongside processes of datafication – mark out their capacities and life-chances as a priori 'different' (Gagen, 2015). Second, the presence of robots – as with other IoT technologies – signals the importance of materialities that look like 'objects', even as in other aspects of technological development the importance of objects may recede (Plowman, 2019). In other words, the inclusive capacities of QTrobot – not least its 'cuteness' – are afforded by its taking the form *of* a robot, rather than a human. The interaction between a screen and a robot may – as a result of the connectedness of both technologies – bear certain similarities that can be appreciated via Ash et al.'s conceptualisation of unit/vibration/tone. Yet robots are not smartphones or tablets, and the very particular kinds of (partial) encounter with robots-as-objects serve as a reminder of the proliferation of material forms that digital technologies can take as they become (part of) media-machines (Bryant, 2014).

Conclusion

This chapter has, building upon Chapter 4, sought to interrogate how *material* stuff about childhoods is always-already enmeshed in and constitutive of media-machines. The specific nature of that meshing – and the extent to which those media-machines are *digitally suffused* – varies enormously, although the example of tyre-burning in Aleppo reminds us that even in some of the world's most precarious places, social media may have a part to play in the circulation of stuff about childhoods. The purpose of this chapter has been to develop modes of thinking and doing, after childhood, which are attuned to the intensifying circulation, visibility and volume of images, texts and material objects that are becoming evident through cultures of digitalisation (and especially social media). Following Bryant (2014), Gallagher (2019) and others, the distinction between media and materialities has been hard to sustain, even if, as I argued at the beginning of the chapter, work on children and *media* and on children and *materialities* has tended to proceed separately. Yet, at a fundamental level, as I have shown, the ever-intensifying circulation of stuff about childhoods works by a fairly smooth (if not seamless) integration of vast social networks around particular child-related hashtags, the massification of images and text, the ability to buy, sell and simply make-visible millions of child-related toys and paraphernalia, and the (sometimes exclusionary) capacities of digital media as they are experienced at the interface (Ash et al., 2018).

It goes without saying that different children, in different places, are positioned differently within such cultures – and that, as I argued at the end of the chapter, even attempts to include children marginalised from other aspects of our (digital) societies may subject them to other forms of marginalisation, stigma or exclusion. With this in mind, this chapter does not constitute a universalising theory or set of theories for considering the circulation and visualisation of childhoods (and children). Rather, I have sought to put forward a range of key conceptual terms that might help to explain some of the ways in which stuff circulates, about childhood, and how that stuff can come to matter in vastly different ways depending on how it/they is/are composed (the example of aeroplane/smoke interfaces being a case in point). Once again, these terms are held together by an impulse towards the 'pull focus' – towards arts of (not) noticing how children and childhoods are both decentred and recentred; how they may be peripheral to any given story, only to (re)emerge even more powerfully as a result; or how a consideration of digital cultures that somehow involve childhood (such as toys excavated through our archaeologies of an online selling site) might actually involve analyses that pay little attention to children's voices or agency – even if, as I demonstrate in relation to the Zapf doll above, or the cars discussed in Chapter 6, a consideration of the infra-generational positioning of toys might enable rather different expressions of children's agency to emerge in the 'damage' done to those objects.

In summary, the key terms introduced and collected in this chapter might afford a range of ways for thinking with media, machines, materialities and other stuff *circulating* about childhoods – and therefore extend the repertoire of ways in which it might be possible to think and do, *after childhood*. These terms have centred on how (social) media, datafication, AI and other digital technologies and cultures may make childhoods and children visible in new or intensifying ways, whilst, simultaneously, hiding or making less visible other aspects of what Livingstone (2019) terms their 'real' selves. I have foregrounded the importance of *platforms* as key infrastructures for the circulation of data (in and beyond schools), images, social networks and objects like toys. In turn, platforms may be specified through the modalities of *control, friction, exposure* and *inclusion* that channel people, data, objects and other flows. I have explored languages for considering *networks* – whether divided, unified, fragmented, clustered or hub-and-spoke (Himelboim et al., 2017). Finally, I have considered the multiple ways in which *interfaces* involving humans, digital technologies, and more, may supplement the ways in which we could articulate the *massification* of stuff that circulates about childhood, and which I introduced in the previous chapter.

I will return to some of these terms, albeit in different contexts, throughout the book. I want to close this chapter by observing that, although these terms might have conceptual and analytical purchase, they are involved in, and not distinct from, stuff that *matters* to children. Beginning where I started: the example of the tyre-burning children of Aleppo demonstrates that we need to attend, critically and carefully to the *specificity* of stuff circulating about childhoods, in different, multiply imbricated registers (whether the tweets or the smoke released through children's own agency). This means, as I have argued many times before, not ignoring children's voice or agency, or representations or 'constructions' of childhoods, but ensuring that they are co-implicated in analyses that can draw out some of the intractability of situations like the Syrian conflict – which profoundly affects but also goes *beyond* children (Kraftl, 2013a). Of course, other stuff matters, to differing degrees, in differing ways and to differing ends: from attempts to 'include' neuro-atypical children through robots to efforts by UNICEF and others to develop international frameworks for children's rights and responsibilities when it comes to 'Generation AI' (www.unicef.org/innovation/stories/generation-ai). Yet languages of visibility, massification and interface could offer a helpful toolkit for analysing how – more than ever in digitally mediated contexts – stuff is circulating about childhoods in ways that matter, profoundly, and whose implications we have only begun to understand. Chapter 8 – which focuses on the lives of plastics and other stuff – picks up this argument in a rather different context again.

6   Infra-generations

After-lives, or, what lies beneath

Every day, global warming burns the skin on the back of my neck, making me itch with physical discomfort and inner anxiety. Evolution unfolds in my genome as my cells divide and mutate, as my body clones itself, as one of my sperm cells mixes it up with an egg. As I reach for the iPhone charger plugged into the dashboard, I reach into evolution, into the *extended phenotype* that doesn't stop at the edge of my skin but continues into all the spaces my human-ness has colonized.

(Morton, 2013: 27)

You would think the media and every one of our leaders would be talking about nothing else. But they never even mention [climate change]. Nor does anyone ever mention the greenhouse gases already locked in the system. Nor that air pollution is hiding some warming; so that, when we stop burning fossil fuels, we already have an extra level of warming – perhaps as high as 0.5 to 1.1 °Celsius. Further-more, does hardly anyone speak about the fact that we are in the midst of the sixth mass extinction: With up to 200 species going extinct every single day. That the extinction rate is today between 1000 and 10 000 times higher than what is seen as normal. Nor does hardly anyone ever speak about the aspect of equity or climate justice [. . .]

If I live to be 100, I will be alive in the year 2103. When you think about the future today, you don't think beyond the year 2050. By then I will, in the best case, not even have lived half of my life. What happens next? In the year 2078, I will celebrate my 75th birthday. If I have children or grandchildren, maybe they will spend that day with me. Maybe they will ask me about you, the people who were around back in 2018. Maybe they will ask why you didn't do anything while there still was time to act. What we do or don't do right now, will affect my entire life and the lives of my children and grandchildren. What we do or don't do right now, me and my generation can't undo in the future. So, when school started in August of this year, I decided that this was enough. I set myself down on the ground outside the Swedish parliament. I school-striked for the climate.

(Greta Thunberg; transcript of TEDx talk, Stockholm, 24th November 2018; https://blog.wozukunft.de/2018/12/19/greta-thunberg-tedx-2018-11-24/#2018-11-24en)

Why generations matter – and how *else* generations might matter

What links the two quotations that begin this chapter? The first is taken from a philosophical text and offers a keen insight into the non-locality and phasing of the hyperobject 'climate change', which I introduced back in Chapter 4. The second was delivered by a (then) 15-year-old schoolgirl, widely regarded as the instigator of the #climatestrike movement that began in summer 2018, and is a powerful example of youth 'voice' and social action. But, both quotations are stylistically arresting; both deploy different degrees of indignation with what humans (or, rather, some quite particular groups of humans) have done to the earth; and both reach into the future – knowingly – beyond the immediate temporal frames of much climate change discourse. It is this latter point that is the crux of this chapter. In their very different ways, and without necessarily using the word, Morton and Thunberg are talking about *generations*. But they are not (quite) doing so in the ways in which we use the term in everyday speech, nor, I think, as do the majority of social scientists who have analysed the term, particularly in recent work on *inter*generational relations (see Chapter 2). Rather, Morton and Thunberg offer starting points for what I term *infra*-generational relations.

I will return to the ways in which OOO theorists might help us to think infra-generationally in later parts of this chapter. For now, before suggesting an initial definition of the term 'infra-generational relations', it is worth stopping to pause a little more on the #climatestrike movement, which was still gathering momentum as I wrote this chapter. As part of the Plastic Childhoods project, we analysed a range of Twitter hashtags that extended both to more generic stuff *circulating* about childhood (see Chapter 5) and to events, phenomena or current affairs that were trending amongst users. Figures 6.1 and 6.2 visualise a retweet network analysis for the hashtag #climatestrike between 11 February and 7 March 2019. Figure 6.3 is a graphical depiction of the same data over time. The latter shows a marked increase in the density of both original tweets and retweets (with a slight time-lag) on and around Friday, 15 February – when school-strikes were staged around the world. Figure 6.1 illustrates the sheer density of network activity around the hashtag to a granular scale, clearly showing a fairly large number of users (each blue dot) towards the centre of the figure being entangled in manifold and multidirectional activity. Meanwhile, figuratively (although not necessarily literally), the 'periphery' of Figure 6.1 shows thousands of user accounts that were less densely and intimately connected with one another – but, nonetheless, that were active around the #climatestrike hashtag during the same period. Figure 6.2 offers a slightly 'cleaner' and less complex view of the same data; here, individual users are more obvious and are more obviously influential. Each line indicates a retweet, but in the case of some users (such as Greta Thunberg), the sheer number of retweets displays as a shaded zone around their account.

I provided an initial conceptualisation of these data – based upon a fairly simple categorical framework – in the previous chapter. Here, though, I want to argue that these kinds of electronic traces are constitutive of a *generation* of sorts. Yet, as I argue in a moment, more traditional Mannheimian notions of 'generation'

#climatestrike (n > 1)

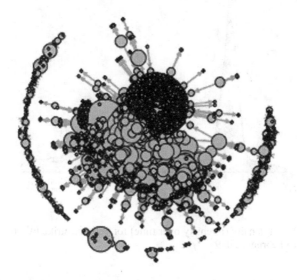

Figure 6.1 Hashtag network analysis for #climatestrike (all tweets)

#climatestrike (n > 10)

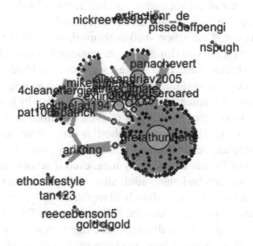

Figure 6.2 Hashtag network analysis for #climatestrike (retweets ≥ 10)

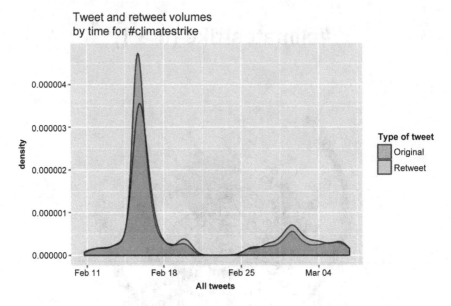

Figure 6.3 Tweet density (intensity over time) for #climatestrike tweets, on and around 15 February 2019

fail to capture the kinds of generation-ing (and the generative possibilities) that are embodied in these digital networks. To some extent, Figures 6.1–6.3 are not visualisations of a traditional 'generation' at all: rather, they are graphic render-ings of *events* that constitute and are constituted by another of Mannheim's less well-used terms: *cohorts* (Brown and Kraftl, 2019). These kinds of patterns were repeated each week and each month – gaining in size, density and geographical extensiveness over time – and were in turn articulated with the marches and other actions that constituted the school-strikes themselves. I use the term 'cohort' as a holding concept for now to denote the 'throwntogetherness' (Massey, 2005) of these tweets, retweets and users; to evoke a sense of individuals and groups "who are – perhaps through an institutional space, club, voluntary organisation or work-place [or event or communicative medium] – tied together by a mixture of chance and design, and who experience facets of their identities *as* a cohort" (Brown and Kraftl, 2019: 616). The notion of 'cohort' is useful here because it does not, specifically, tie into any readily categorisable identity position. Perhaps most sig-nificantly, not all of Twitter's users are young, even if school-strikers themselves were; the school-strikers had many adult allies, both on and beyond the medium of Twitter. Rather, we witness a cohort held together – however tenuously – by an issue and, in a sense, by a digital marker: #climatechange. Thus, without repeating the arguments of Chapter 5, the diagrammatic representations of Twitter activity around #climatestrikes denote at least two kinds of decentring: of those individuals/ groups labelled 'young' or 'children' and of expectations about what kinds of individuation (or 'agency') and what kinds of relationality might be required for a social network such as that in Figure 6.1 to emerge, as human 'actors' (and cohorts) are constituted with, in and as 'media' (Bryant, 2014). Indeed, beyond its figurative or representational capacities, what is the status of this social media network? Is it a (hyper)object, especially when seen as one surface of the phenomenon of school-strikes?

But let us not get too carried away (yet). For, whatever the symbolic or affective *powers* of these media-machines (Bryant, 2014), I want to hold those powers alongside an acknowledgement of the *generational* politics and sensibilities that Morton and, especially, Thunberg articulate at the beginning of this chapter. The notion of cohortness might help access the *internal* workings of #climatestrikes, but it does not tell us everything we need to know about the imperatives that drive them.

A look at the *content*[1] of some of the many thousands of tweets under the #climatestrike hashtag alone provides a sense of these generational politics and sensibilities. Of just over 205,097 tweets during the period 11 February–7 March 2019, a not insubstantial 5,215 (2.5%) directly referenced the term 'generation'. A very small selection of those tweets evidences the observation made above that the loose 'cohort' of users cohering around the #climatestrike hashtag are both younger and older.

> there's real hope in that activism. Takes me back to my student marches! The next generation cares [clapping emoji]

> [username] is done waiting for adults to save her generation from climate change and we are HERE [clapping emoji] FOR [clapping emoji] IT [clapping emoji]

> We are the first generation that knows we are destroying the world, and could be the last that can do anything about it

> This is really the #ClimateGeneration we are talking about here

> Climate crisis and a betrayed generation

> My generation will have to live in a #climate changed world

> I want my generation's voice to be heard and listened to. We stand to lose the most from catastrophic climate change

> My generation trashed the planet. So I salute the children striking back . . . those of us who have long been engaged in this struggle will not abandon you

A quantitative content analysis of the same dataset reveals a considerably higher proportion of tweets, including the terms 'future' (40,659; 19.8%), 'hope' (12,137; 5.9%), 'child' (30.922; 15/1%) and either 'care' or 'save' (10,256; 5%). But whilst a proportion of those tweets could be interpreted to be positive (even if some might be concerned for the 'future' or reference lost 'hope'), many are concerned with more overtly negative projections of the future. A total of 20,114 (9.8%) reference the terms 'extinction' (4,220), 'devastation' (3,216), 'damage' (654), 'threat' (2, 458), 'crisis' (3,299) or 'emergency' (6,267). Each of these terms articulates different affective dispositions to the future (Anderson, 2010); but, taken together, they question the commonplace ways in which children (as a future generation) are discursively and affectively connected with the notion of future-as-hope (Kraftl, 2008), or the tactics through which green thinking activates a particular 'timescape' by emphasising the implications of global warming for future generations (White, 2017). Greta Thunberg offers a different cue, however, rooted not in hope but a realisation of a looming crisis that requires *action*:

> And yes, we do need hope. Of course, we do. But the one thing we need more than hope is action. Once we start to act, hope is everywhere. So instead of looking for hope, look for action. Then and only then, hope will come today.
> (Greta Thunberg; transcript of TEDx talk, Stockholm, 24th November 2018; https://blog.wozukunft.de/2018/12/19/greta-thunberg-tedx-2018-11-24/#2018-11-24en)

Therefore, and resonating with the ethos behind my use of the term 'after', I argue that we need, to some extent, to keep hold of notions of 'generation' and 'intergenerational relations' but that they need to be fairly radically revised – as *infra-generational relations* – in order to explicate the more-than-generational and more-than-human processes, energies, materialities, technologies, media, affects and more that constitute the temporalities in which childhoods are enveloped, and of which they are productive. To do so is not merely to enable an analysis of the different temporal and spatial horizons implied by the quotations from Morton and Thunberg that began this chapter but to attend – as per materialist-inspired philosophies – to the more-than-human constituents of those timings and spacings as 'generations' of different sorts. In other words, and building on the concepts of circulation and massification that were introduced in the previous chapter, and of resource-power and materiality developed in the two chapters before that, this chapter considers in greater depth the spacings and especially *timings* involved in thinking and doing, after childhood.

From generations and intergenerational relations to *infra-generations*

As I argued in Chapter 2, a fairly long-standing and familiar – if less common – route for academics wishing to address the centrality of (individuated) children within studies of childhood and age has been to think about age relationally: to think of child–adult relations (Alanen and Mayall, 2001) or intergenerational relations (Hopkins and Pain, 2007). Rooted in Karl Mannheim's conception of the term, youth is seen as a phase of generation-building, which coheres around political events (in particular), technologies and other aspects of shared experience and generational self-awareness (Burnett, 2016; Worth, 2018; Brown and Kraftl, 2019). Although influential, in the past couple of decades, a range of scholars has sought to extend Mannheim's theorisation of generations in some important ways. For instance, arguing that Mannheim's focus on politics and synchronicity in the birth-order were limiting, it is now evident that generations and their *experience* are mutable and iterative – constituted via mundane, everyday, embodied and emotional acts, in combination with membership of (post-)subcultures (Woodman and Wyn, 2014), or via the differential take-up of media or technologies both within and between (re)generations, such that arguments about 'digital natives' are hard to sustain (Wachelder, 2019). Elsewhere, powerful critiques of the assumed subject-positions of those designated as 'child' or 'adult' have stressed relations of *inter*-dependency, where, for instance, acts of care *by* children *for* adults have undermined commonplace understandings of the apparent verticality of the generational order (Punch, 2002; Evans and Thomas, 2009).

Bearing particularly in mind these extensions of Mannheim's original work, the study of generations is a potentially potent way to decentre children within social-scientific analyses – particularly in response to what Hopkins and Pain (2007) identify as a disproportionate focus upon the young and, to a lesser extent, the old in work on *age*. Indeed, there are plenty of examples of excellent work, particularly on inter- and intra-generational relations (the latter referring to dis/congruencies *within* identifiable generations; see Lulle, 2018). Yet the notion of 'generation' has been subject to many critiques (e.g. Horton and Kraftl, 2008). Notably, just a couple of years ago, Woodman and Bennett (2016: 36) were able to claim that social-scientific work on generations was (still) "returning to prominence" but that the term can (still) be "criticized in particular for conceptual fuzziness, for obscuring intra-generational differences and inequalities, particularly by class, and ignoring intergenerational solidarities". This is a critique of the persistence of "generationalism" or a simplified and exaggerated picture of generations, which dates back to early twentieth-century European intellectuals and which can still be found in today's popular discourses as well as in academic studies (Purhonen, 2016: 94). Incidentally, Purhonen's (2016) response is to turn to Bourdieu's work on social groupings and classification; others (like Woodman and Wyn, 2014) also draw on Bourdieu, alongside Beck and Giddens, in their attempts to draw together questions of inequality, change and socio-cultural variation (Hoolachan and McKee, 2019).

Yet, bearing in mind the opening section of this chapter, and the chapters that have preceded it, it is my sense that the notion of generations could do *more*. The term can be pushed far further, without, as I argued above, dispensing with it altogether, because I firmly believe that – as the opening section of this chapter also demonstrates – it has significant *political* purchase, if nothing else. Despite a temptation to replace the term with others – like temporality or chronology – the notion of generation offers a kind of specificity that could be vital in thinking and doing, after childhood.

Therefore, in the rest of this chapter, I argue for, and exemplify, a concept of *infra-generations* that performs two main tasks, and which requires the prefix 'infra-' – at least at this juncture – to signal the intent and import of those tasks. First, as far as I am aware (despite much recent work in this vein on *childhood*), for the first time, it offers a systematic theorisation of *more-than-social* generations. Pushing somewhat the definition of the term, I speculate about the 'generations' of objects, like toys; of earthly systems, such as weather, climate, sedimentation or sea-level rise; and of demographic trends and the evolution of the human species (and questions of life and death to which I return in Chapter 9). Second, also, as far as I am aware, for the first time, I place generations within a much longer historical purview. In particular, in combination with original analyses from the Plastic Childhoods project, I work with and re-read recent work on the archaeologies of childhood, which extends back (in some cases) far further in time than even the more historical studies known to most childhood scholars.

The choice of the prefix 'infra-' is strategic, for several reasons. I dismissed 'extra' or 'super' because they felt somehow arrogant and dismissive of previous work on generations, affording some kind of sensibility of being 'above' or 'beyond' that would, in particular, be counter to the ways in which critical race and queer theories thread through this book in ways that mitigate against forms of grand claims-making. I also toyed with the prefix 'intra' (as in Barad's [2007] intra-action) but, as indicated above, the notion of 'intra-generational' relations has some specific connotations. Instead (a bit like the term 'after'), the prefix 'infra-' denotes something *below* or *beneath*; something *secondary* rather than originary (and therefore definitely *not* structural or foundational); something perhaps *slower* (as in the term 'infra-sonic'); something perceived as *inferior*, perhaps marginalised (as children often are); that which sits *behind* (rather than beyond, as in many dispositions to the future); and, crucially, somehow *after* – even if, as I will show, in chronological terms, infra-generations refer to that which predates the present, often by many centuries.

My prefixing of generations with 'infra' is also inspired by Latour, via Hultman and Lenz-Taguchi (2010: 536), in accounts of post-qualitative, post-human forms of reflection:

> [w]e can never reflect upon something on our own; to reflect means to *interconnect* with something. This corresponds to Latour's concept of *infra-reflection* that takes into account that reflection is always done in the midst of a complex network and thus immanent to a wide variety of forces and never the product of an isolated individual that reflects upon something from an external point of view (see Latour, 1988). Thinking is not something that is grounded on a decision or a rational cataloguing of different external objects: rather, it is an event that happens *to* us – it 'hits us' or 'invades us'.

A question guiding this chapter, then, is what might happen if we replace the word 'reflection' in Hultman and Lenz-Taguchi's words with 'generation'? What might be the conceptual, political and methodological effects and affects of such a move? Certainly, this would be to enact some kind of Deleuzian reading of generations as emergent and becoming, as much as or as well as childhoods are becoming or going on (Worth, 2009; Horton and Kraftl, 2006b). But, for me, more importantly, the quote above signals a form of imm*a*nance (note, not imminence) that is both beyond/beneath the individual but not 'grounded' on a view of the more-than-human as somehow external-to, or resources-for, human action. Building on previous chapters, this is a provocation and a political act as well as a conceptual one: it is to foreground those aspects of what are or might be 'child-hoods' and, especially, 'generations' that have routinely gone ignored in chief social-scientific accounts of children and young people; it is to call out those facets of childhood that have quite simply been *thrown away* – including skeletal remains and the artefactual traces of long-gone childhoods; and it is to literally and figuratively scratch the surface of OOO – to explore *what lies beneath* when we encounter the *after-lives* of children's objects and bodies, and when those objects and bodies are fragmented, shattered, broken, scratched, only to be dis-covered later in archaeological assemblages. In a chronological sense, this chapter is partly about finding ways to recognise children many years after they not only become adult (and no longer child) but *after they have lived*. In a less straightfor-ward sense, and inspired by Meillassoux (2010), it is a recognition of childhoods *anterior* to even the modern notion of childhood that is the frame of reference for most social-scientific scholarship – because, in a slightly odd and perhaps macabre twist of Meillassoux's argument (which looks *after human finitude*), the material-cultural and skeletal remains of Roman and even earlier childhoods are evidence of (human) childhoods that are, ostensibly, anterior to our contemporary phenomenological horizons.

In Chapters 3 and 4 I argued for arts of *not* noticing (following Tsing), and the 'pull focus' (following Hitchcock and Morton), as strategies for apprehend-ing the impulse to decentre childhood (Spyrou, 2017). In response, this chapter focuses on three ways to *find children* (and childhoods) in historical and archaeo-logical records. This phrase is chosen deliberately to reflect over two decades' worth of work by archaeologists that not only promotes the study of children and childhoods but which has, explicitly, sought to develop innovative methodologies for finding (Crawford, 2009) and discovering (Ember and Cunnar, 2015) them amongst archaeological and archival assemblages. As Álvarez (2017: 123) suc-cinctly asks: "how are children manifest archaeologically?" Although explicitly mirroring social-scientific approaches to childhood – for instance, in attempts to uncover children's agency – archaeologies of childhood also offer ways to chal-lenge, decentre and extend those approaches. They do so through what I term the 'after-lives' of children and childhoods, exemplified by a key paradox in archaeo-logical work on toys (Crawford, 2009): that is, that despite a focus on children, one must always start with assemblages containing a vast array of stuff – stuff that is usually classified according to *adult* function (e.g. weaving or cooking) but which, once classified as a plaything, is often discarded and deemed insignifi-cant. Therefore, any attempts to 'find' children must always be characterised by a willingness not (only) to *start with* children – and the almost uniform absence of that mainstay of childhood studies, children's 'voice', makes this a particu-lar challenge. Rather, 'finding' children requires a painstaking (re)construction and collaging of evidence from multiple sources, an open, generous and creative mindset that enables inquiry into how particular objects might, after all, have been 'toys' or other traces of childhood (and neither 'adult' objects nor simply 'waste'), and a fastidious approach to the very physicality of objects themselves – to where they were deposited, and how, to how they have been worn, or scratched, to how bones have been marked by the wear and tear of lives lived and the ongoingness of lively processes after death. In this analysis, the notion of the 'cut' – first introduced in Chapter 3 – re-emerges and figures highly as an analytical device. Thus, drawing on original findings from the Plastic Childhoods project, and upon a critical, synthetic and creative re-reading of historical and (especially) archaeological literatures on childhood, this chapter outlines three 'cuts' through which it might be possible to 'find' children and childhoods: through after-lives of 'toys', of an online selling platform and of bones.

Finding children I: after-lives of 'toys'

Let us, with the greatest respect and care, ignore children just for a moment. This, I hope, will enable us to return to children in all-the-more instructive, evocative and, quite possibly, poignant ways. With an eye firmly on theories of materiality, consider *assemblages*. In archaeological terms, assemblages are sets of objects found together – often, but not only, uncovered during a dig event – that are often taken to denote synchronicity and therefore correspondence with a particular culture or community. Depending on their age and position, assemblages could have been subject to multiple forces, such as deposition by humans, whether in a single moment of divestment or the accretion of materials over decades; subsequent disturbance or damage by humans, other animals or plant roots; and weathering and other forms of perturbation, caused by a range of factors from soil creep and cracking to leaching. Like the doll in the apartment in Luz and its many companions, assemblages are not, then, simply 'human' artefacts but are constituted by both more-than-human material stuff and more-than-human processes.

Although (from my reading) archaeologists seldom need to have recourse to philosophical renderings of the term assemblage, there are resonances. In Deleuze and Guattari (1988: 16), for instance, we find reference to constellations that resemble a 'fragmentary whole'; they are not unitary bodies but, rather, "machines, defined solely by their external relations of composition, mixture and aggregation" (Nail, 2017: 23; see also Bryant, 2014, and Chapter 3 of this book). Counter the risk that *any* collection could be an assemblage, it is the fragility and 'provisionality' of an assemblage, energised by forces of 'gathering, coherence and dispersion' that characterises its emergence (Anderson and McFarlane, 2011: 125 & 124; also DeLanda, 2019).

Yet *archaeological* assemblages are also somewhat dissonant from these philosophical deployments of the term, in some important and potentially generative ways. For instance, Nail (2017) argues that assemblage-thinking (like ANT) looks beyond individual elements of an assemblage, and beyond essences, to relations: to ask not what is this individual 'thing' but how did this constellation come to appear as such, when and through which agents (or 'personae'). Whilst the very idea of an archaeological assemblage also requires an emphasis on relations – after all, it is the *collection* of stuff that is taken to be indicative of a particular culture – as we shall see, archaeologists are also interested in the properties of *individual* objects, even if themselves analysed relationally. This observation resonates with my analyses of individual and massified objects found on the online selling platform in the previous chapter, and I will take up this point again in relation to OOO, later on.

Meanwhile, many assemblage-thinkers consider that – if not erroneous – any move to consider assemblages solely as a noun (as 'a constellation') rather than a verb (always 'constellate-*ing*' or 'disperse-ing') is potentially problematic because it effaces the properties of emergence (Anderson and McFarlane, 2011). Yet, again, although evidently cognisant of the processes that created an assemblage, the very viscerality of materials found in a 'cut' or other archaeological formation and the very process of uncovering and documenting those materials *as a collection* seem also to imbue them with a kind of ontological stasis and boundedness. As Cessford (2017: 164) has it, an assemblage denotes "a group of artefacts found in close con-textual association". Thus, and finally – other than, perhaps, the notable exception of Deleuze and Guattari's (1988) brief nod to celestial constellations – the time-frames under consideration here are simply longer than those attended to in most attempts to empirically examine assemblages. These points about temporality are important because they put any notion of childhood – and questions of children's putative agency – within a spatio-temporal tension or paradox between the appar-ent stasis of an assemblage encountered *now* and the conditions that led to that assemblage being deposited as such. In other words, these are the spatio-temporal complexities and juxtapositions of the after-lives of assemblages that may or may . not contain traces of childhoods.

Take the example of large domestic assemblages deposited in what is now the centre of Cambridge (UK) since 1750 (Cessford, 2017). Cessford focuses on 'cuts' – features like a ditch or cesspit, literally cut into the earth, filled with household waste and, often, subsequently covered over. These assemblages therefore appear to be closed groups, which might tell us about (for instance) the socio-economic circumstances of the family that deposited them. Thinking back to Chapter 3 for a moment, these 'cuts' are other local manifestations of what I termed 'cuts through the earth': smaller, earlier instances of waste deposition that nonetheless share something with the eco-park in Guaratinguetá. They are all part of the same hyperobject (Morton, 2013): the becoming-earth of human detri-tus (or, 'Trash'). These cuts are, too, redolent of assemblage-thinking: of "(1) 'a selection cut' allowing something to pass through and circulate [crudely, the act(s) of deposition] (2) 'a detachment cut' that blocks part of that circulation [crudely, the act of covering over]" (Nail, 2017: 29).

Yet, as I argued above, whilst Cessford (2017: 175) recognises "that the over-all assemblage biography is at least potentially greater than the sum of its con-stituent parts", he nonetheless devotes significant attention to the questionable status of individual 'items' within the Cambridge assemblages. Bearing in mind how OOO deals with objects, this discussion is both fascinating and instructive. Cessford (2017: 169–170) makes several points in this regard: that what con-stitutes an 'item' under a contemporary viewpoint might not have done in the past; that, as a result, constituent components of an 'item' (such as a teapot and lid) might not have been conceived as such, where if one broke the other might have been re-used in conjunction with a different pot or lid; where 'items' might be broken – whether in the course of their use, in the act of deposition, in order to make them easier to transport for deposition, or after burial; and, where – as in some Cambridge cuts – assemblages contain materials from midden heaps (in the form of soil/rubble/broken ceramics) that may have been used to backfill or compact waste, but which stuff does not constitute a haeccetic, bounded 'object' although it is nevertheless *matter* (Horton and Kraftl, 2018).

In other words, Cessford offers a series of important ways to question the item-ness (or object-ness) of artefacts that hold in tension the status of individual items with, in and as the other constituents of the assemblage (and hence the assemblage's relational and chronological composition). Thus, it could be posited that such observations alone expose the historical specificity of modern ways of conceiving objects. These modern conceptions could be thought of as (infra-)generations; or, rather, a 'modern' generation of objects whose object-hood is made to appear rather more ephemeral than solid, in light both of the kinds of archaeological matter found in Cambridge and the ways in which objects are becoming re-cast and re-thought in emerging digital cultures (see Chapter 5). Moreover, to be provocative, perhaps (as a result) they also expose the limitations of the ways in which OOO, new materialist and other contemporary theorists articulate the object-hood of objects. Whilst a proper examination of this provocation is beyond the scope of this book, I offer a partial response by returning to the issue of breakage later in this section.

Cessford makes a further important distinction that offers a signpost to something of what I want to argue in terms of infra-generations. The distinction is between three different – but overlapping – forms of temporalities. First, he sees the cuts in Cambridge as 'materialized temporalities' – or '"temporal palimpsests', where items of different ages and life spans are the constituents of a single depositional episode" (Cessford, 2017: 175). In other words, even if the objects themselves were made or used at different times, they were deposited together. At the same time, however, as the example of midden used to backfill cuts demonstrates, there could have been multiple depositional events – alongside the more-than-human processes of soil movement, leaching and action by flora and fauna noted earlier. Thus, each assemblage could, in effect, be constituted by multiple, *multiple* temporalities – of the objects themselves and their individuated 'biographies' (Crawford, 2009; something to which I will turn below), of their amalgamation in individual depositional events and of their entanglement with/in later depositional events and other processes. Consequently, as Dixon et al. (2018: 117) provocatively put it, cities – like Cambridge – can be viewed as 'emerging landforms' in the Anthropocene, and especially through conditions of degradation, neglect and wastage that become entangled with geological or geomorphological processes – however apparently minor. But put loosely at this juncture, I want to ask whether we can see these kinds of entangled processes as forms of *generation*: both as generative (of the assemblage and of a series of pertinent questions about items) and as generational (as in of more-than-human timescales that nonetheless articulate human generations).

If in this rendering of 'cuts', humans – and human *children* – appear to have been somewhat peripheral so far, then on my part this is intentional. It is only in this way, I think, that we can reconceptualise generations as *infra-generations*. How, then, are *human* generations articulated in the Cambridge assemblage? Cessford makes an important observation in linking – conceptually – the generation(s) of the assemblage with *human* lifetimes. He notes, for instance, that it would be important to map alongside the stratification of the Cambridge domestic assemblages the fortunes of the families that deposited them. He asks whether, for instance, depositional events represent the changing social or economic fortunes of a family (whether improving or declining); or whether they mark particular lifecourse phases that are intimately connected with generations – birth, ageing, dying or moving house; or, indeed, if they represent – in the most traditional sense of the word – generational shifts in perceptions of the utility or worth of objects (Cessford, 2017). Critically, again, - and like the doll found in the Luz apartment - especially without accompanying documentary or oral-historical evidence, it may be very hard to draw out correlations between the fortunes of a household, or *particular* moments of human agency, or *particular* conceptions of objects. In passing, then, notwithstanding the likely entanglement of the multiple, multiple temporalities of assemblages with/in/as human generations, we once more run up against the limitations of correlationism, as the advocates of OOO frame them (see Chapter 1).

The issue that remains, then, is one of 'finding' children in the Cambridge 'cuts'. Cessford's (2017) work is so interesting to me not because children are *not* mentioned but because they are mentioned only *once*. He focuses upon a sub-section of items that seem to have been deposited following the death of an individual – a Sarah Hopkins – in 1843. These include "a jug, a child's cup, and a miniature watering can" (Cessford, 2017: 185) – noting here that miniature versions of 'ordinary' adult objects are often taken as evidence of children's presence at a site (Harper, 2018). These items were likely to have been important mementoes for Sarah, either of her own childhood or that of her children, and therefore held emotional significance. Notwithstanding the significance of these traces of childhood within Sarah's life and eventual death – and hence her treatment within generations of the same family – it is the way in which these items are narrated in a caption accompanying an image thereof that is even more fascinating. Here, the emotional significance of the items is couched in terms of the ways in which objects in the assemblage relate to one another (cf. Harman, 2010b) and constitute part of the stratigraphy of the site (cf. Dixon et al., 2018). To quote at length from the caption:

> [s]ection and photograph of Cellar 4 at the Grand Arcade site, Cambridge, backfilled 1843–45, demonstrating the different quantities of 'items' in different deposits plus the degree of inter-connectedness of the material from different stratigraphic deposits. There are also a jug, child's cup and watering can interpreted as late 18th-century personal items, their presence serves to underlie the connection of the deposition of this material to the death of Sarah

Hopkins and the issue of the potential emotional involvement of those who discarded the material.

(Cessford, 2017: 176)

Biographies of objects

In the rest of this section I want to turn to studies that are more directly concerned with childhood. Yet my starting point is, once again, the observation that doing so may necessitate somewhat of a focus away from childhood – or, if not from childhood, then from established tropes for studying children. Crawford's (2009) work on the 'biography' of objects is instructive. She focuses on a site at Mucking, Essex, inhabited between the fifth and eighth centuries AD. A guiding problem for Crawford's work is why – despite other evidence that children lived at the site – there appears in the archaeological record not one example of either children's material cultures generally or of toys specifically. Noting that play is more-or-less a human universal, and something that is often taken to differentiate children from adults (although see Woodyer, 2012), then objects that could ostensibly have been played-with (what we now term 'toys') would be an important trace of childhood. The problem – or the paradox – here is that anything that cannot categorically be classified *as* a toy is immediately classified as a non-toy – and therefore an artefact of *adult* life.

Crawford's (2009) ingenious solution is – in a sense resonating with theories of assemblage – to move away from toys as *objects* to a livelier sense of toys as *concepts* – with the important corollary that any-thing could be(come) a plaything. This precisely entails a decentring *and* recentring of children – an example of the 'pull focus'. Rather than (only) look to the biographies and lifecourses of humans, Crawford asks whether we can theorise a 'toy stage' in the biography or 'lifecourse of objects' (Crawford, 2009: 55 & 57). Akin to – but rather differently from – ANT and new materialisms, this approach *starts with objects*, and not children – although of course it does so precisely to 'find' children and to attempt to say something all the more significant about them. In the absence at most sites – and certainly at Mucking – of documentary evidence about these objects' biographies, Crawford (2009: 62) notes that a starting point could be the 'depositional pathways' of objects – in a sense, the (infra-)*generations* of objects. She notes that, in many cultures,

> objects enter a stage of use as a toy at the *end point* of use in the adult world, and prior to final abandonment. A broken, worn or unvalued object may be deliberately passed on to a child: in this sense, children may function in the biography of an object as a temporary or transitional phase in the deposition process.
>
> (Crawford, 2009: 62)

This is a striking passage, for several reasons, and not only for the 'pull focus' manoeuvre in which children are re-placed with/in the biography of an object, rather than the more familiar move of making an object (a toy) a part of a child's lifecourse (although that is *also* implied here). Rather, it is striking because of the chronological positioning of childhood *after* adulthood – the penultimate 'generation' in the biography of an object lies in its association with a human generation that is presumed to come *before* adulthood: childhood. This, then, is a fundamental twisting of the notion of generations that is facilitated by a more-than-human analysis of the term, alongside attempts to move beyond teleological accounts of ageing in literatures on childhood (see Horton and Kraftl, 2006b). In terms of thinking and doing *after childhood*, then, it is not only evidence that childhood can come *after adulthood* but that children are key players in the *after-lives* of objects – such as those objects-becoming-toys. Note that – in distinction to some of the online selling platform items that I shall look at a little later – this final stage is not that perhaps more familiar one where an adult might retain, collect or display toys for reasons of 'nostalgia' or some presumed (and adult-inscribed) cultural or economic 'value'. Rather, these items *matter* (Horton, 2010) precisely because of their *loss* of value as understood by adults, their *loss* of utility – and what they therefore *gain* (as a form of what I termed in Chapter 3 'resource-power') as they afford play. Importantly, this is a reversal of the often stultifying ways in which childhoods are memorialised or monetised through toys and again affords a sense that thinking and doing, after childhood, may both enable us to uncover aspects of childhoods that have hitherto been hidden and to recognise other potentially politically progressive ways to undertake childhood scholarship in ways that do not (only) privilege notions like voice or agency (Kraftl, 2013a).

Moreover, the above passage is striking because it points to the after-lives of objects in the sense of their (at least symbolic) 'death' and – without stretching the metaphor too far – revival once uncovered by contemporary archaeologists. Specifically, although adults may have been responsible for the eventual discardment of such items, it is quite conceivable that children themselves deposited materials. It is here that questions of *space* and *materiality* surface – again, in the absence of any reliable ways to gauge children's voice or agency. On the one hand, as Crawford (2009: 63) puts it, "[t]he most likely archaeological evidence that such [. . .] items were ever incorporated into children's play, should any hint survive at all, will be in their location and in their unnatural proximity to other objects". Thus, the microgeographies of deposition – of a possible playfulness not only in how they were used but how they were left – are indicative of their divestment by *children*. What is most notable, as other studies show, is that children tended *not* to leave such items in 'cuts', as per the Cambridge site. Rather – and drawing, not incidentally, upon more contemporary, ethnographic accounts of children's use of outdoor spaces – the 'unnatural' constellations that Crawford references include marbles left on patio spaces at the San Pedro Chohul hacienda on the Yucatán Peninsula, Mexico, in the early twentieth century, spaces which were clean and safe for children to play (Álvarez, 2017); or what Dozier (2016: 65) terms 'child-calculated concentrations' of toys and (especially) plastic items left at the wooded edge of a lot previously abandoned by adults in the town of Shabona Grove, Illinois, in the 1960s (compare Ward, 1978, on children's play in plots left vacant by adults; also Harper, 2018). This kind of observation is exemplary of what I am trying to get at in looking *after* rather than 'beyond' childhood (see Chapter 1). Thus, in Chapter 8, I reassert the importance of including rather than occluding more 'traditional' social-scientific research about children's spatial practices and mobilities in the context of contemporary children's experiences of plastics, metals and other elements.

Significantly, in all of these cases, it is possible to ask questions about how the micro-geographies of 'child deposition practices' (Crawford, 2009: 64) articulate a range of multiply scaled and (inter-)generational, socio-economic and environmental chronologies (what I seek to bring together under the term 'infra-generations'). The case of Shabona Grove is of particular interest. Dozier (2016: 61) starts from the perspective of an interest in children's agency – less in the sense of *social* agency underscored by many social studies of childhood and more in terms of "how children's actions can be seen as a lasting influence on the landscape". Notably, Shabona Grove was a prosperous town until the late nineteenth century, when it was omitted from the first wave (or generation?) of railroad-building. With the railroad instead running through a town just a few miles away, Shabona Grove experienced rapid depopulation as (especially) younger people sought work; the appearance of a spur during a later wave of railroad expansion could not reverse this trend. By 1921, the town had just 61 residents, and most of the town's physical infrastructure had been torn down, although inhabitation continued in some tenuous forms until the 1960s. Thus, many of the child-curated deposits found on the site during recent archaeological research reflect the longer-term, cross-'generational' socio-material rhythms of life itself at Shabona Grove during the first half of the twentieth century. Yes, the plastic toys, tomato ketchup bottles, television tubes were emblematic of the burgeoning mass-production era in the 1960s US, and of what became a distinct generation that grew up with/in mass popular cultures (as [Ghosh, 2019: 277] has it: an "(en)plasticized" generation; see Chapter 8). And yes, second-growth trees have colonised children's depositional sites, their roots and detritus becoming entangled with/in/as those assemblages. As in Cambridge, more-than-human processes may inflect further processes of deposition, covering-over, stratification or what archaeologists term 'taphonomic' processes (Dozier, 2016: 67): the decomposition, sedimentation and/or fossilisation of human artefacts and remains as they literally become earth (a point I return to in the next section). However, the manifestation of these popular cultural media (Bryant, 2014) also evokes other concerns, which were and are highly localised, for Dozier (2016) hypothesises that – in some ways counter the arguments above – we cannot assume that children's efforts to deposit items were solely playful, or playful at all. Rather, they may have engaged in deliberate acts of divestment in attempts to come to terms with the traumatic social and economic stresses of life at Shabona Grove, facilitating some semblance of control – akin to the children who collected warhead shrapnel in London's bombsites during the Second World War or, if this comparison is not too much of a stretch, children's tyre-burning in Aleppo. Whatever the case, these 'child-calculated concentrations' articulated multiple temporalities – generations – at different spatial scales – from railroad investment to mass-production and mediatisation, which *mattered*, profoundly, in ways extending to and beyond 'play'.

In sum, work on the archaeologies of childhood artefacts offers – via an insight into *what lies beneath* – a distinctive set of ways to 'find' childhoods, a distinctive 'cut' through earthly assemblages. Thinking-with the after-lives of toys, in particular, opens out a range of questions about some of the longest-standing concerns of childhood studies – not least about the chronologies of generations, play and children's agency. Through a slightly speculative reading of some key examples, I have also argued that archaeologies of childhood might enable us to think differently, to think creatively, to think more broadly and to think otherwise, about generations that extend both beyond humans and beyond humans' spatial and temporal frames of reference. Indeed, my own speculations have been inspired by styles of thinking and doing in archaeology that, equally, involve measures of creativity, imaginativeness and the making of apparently eclectic or eccentric connections. This is neither to deny the rigour nor the carefulness with which archaeologists do their work; in fact, quite the reverse. For me, the thought and action required for "taking care of country and accounting for generations of entangled human and non-human entities" are vital narratives – 'speculative fabulations' – that express rather than efface the lives of marginalised Others (like children) in the Anthropocene (Haraway, 2011: 5 & 1). Whilst some may in fact seek to question the role of almost *any* scientific method – at least, those of the 'hard' sciences (Stengers, 2015) – I want to ask whether archaeologies of childhood offer starting points for and examples of the kinds of 'radical' interdisciplinarity I call for in thinking and doing after childhood (see, in particular, Chapter 8). Perhaps these are some of the very 'minor stories featuring minor players' that, collectively, we should be telling (Taylor, 2019: 14; also Katz, 1996). Whether or not these narratives might be categorised as such, this section has continued to develop 'arts of (not) noticing' by setting out some initial ways to think childhoods infra-generationally.

Finding children II: after-lives of an online selling platform

As part of the Plastic Childhoods project, we undertook what we termed 'online archaeologies' of a cross-section of childhood-related objects (methodological details can be found in Chapter 1). Our principal source was an online selling platform (the same as the one discussed in the previous chapter), where we focused on objects including toy cars and dolls. Following quantitative (i.e. API scraping) and qualitative (i.e. textual and visual) analysis, we were able to draw out three broad 'categories' of object: 'contemporary selling' (modern pieces whose value lay largely in their relative 'newness' and intactness); 'house clearance' or 'job lots' (often loosely defined collections of toys, such as 'Used job lot of Disney cars items'); and 'vintage' (items ascribed some sort of value – both 'positive' and 'negative' – given their age and physical condition).

Reading and juxtaposing these online selling site items – especially those in the 'house clearance' and 'vintage' categories – alongside those found in physical archaeological 'digs' and 'cuts' is a fascinating endeavour. Every item – however scant the description – has a 'biography' in the sense that Crawford (2009) uses the term. Interestingly, at least in terms of the items we analysed, those biographies rarely (and barely) related to the *people* who had owned or played with the toys. There was, for instance, little mention of older toys having belonged to parents or grandparents or other older family members, and, therefore, the most explicitly 'generational' element of toys' biographies were often missing. Yet if we take the term 'generation' a little more broadly – as I seek to do in this chapter – then other biographical details come to the fore. In particular, and recalling the items found at Shabona Grove, there are plentiful references to the *decade* in which an item was manufactured, the *place* (usually the country), the *materials* and engineering techniques used (for instance, whether diecast metal for cars) and, of course, the make and model. For instance, item 6 in our 'vintage' collection is a 'Ford Cortina and Caravan Plastic Toy Car Jimson Made in Hong Kong 1960s' (Figure 6.4).

Often, as toy collectors will appreciate, it is the combination of these different factors that means a particular item will be ascribed worth – in the case of the Ford Cortina and Caravan, financial worth, because the piece eventually sold for £100. But more importantly, for the purposes of this chapter, these details intersect to offer a sense of the *generationality* of toys. For instance, as at Shabona Grove, the unique admixture of place and time of manufacture is an embodiment of gen-erations of toy-making techniques. Reflecting on the (much earlier) toys found in the England and Wales portable antiquities scheme, Harper (2018) indicates how, thanks to emerging engineering and transport technologies, similar styles of dolls were found in most areas of England, whilst from 1600 onwards, manufactured toys of various kinds were being bought and sold throughout the country. Clearly, these technologies do not operate in isolation – rather, in conjunction with con-temporaneous beliefs and knowledges about child-rearing and education (Crewe and Hadley, 2013), their gradual introduction through the domestic practices of

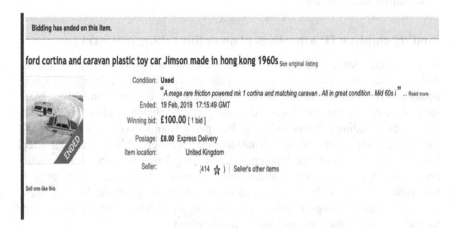

Figure 6.4 Image of Ford Cortina and caravan toy car taken from an online selling platform, 3 April 2019

generations of families (and children) and the fashions or popular cultures of which these toys were a part. Equally, there are generations of toys manufactured in Hong Kong since 1945 – of which (in turn) successive generations of children and adults will be aware and which, as I indicated in Chapter 2, may be accompanied by generational forms of racial claims-making (e.g. Chen, 2011). And there are particular kinds of brands or toys associated with different generations of children – and often intra- or sub-generations or 'cohorts', because those toys may only have been available to a subset of children or for a much shorter time period than a 'generation', as Horton (2010) so beautifully illustrates in his analysis of popular and material cultures associated with the British pop band S Club 7. It is these intersections of place, manufacture, transportation, fashion, mediation, family practices and child agency that – for me – speak of some of the ways in which infra-generational analyses might proceed. It is also worth noting, in passing, that many of these items reverse again the trend seen at Mucking and elsewhere, because the (at present) final stage in an object's biography is precisely to become an object of *adult* interest/collection, after its use by a child (take, for instance, a 'vintage mafco car', item 7 in our 'vintage' collection, about which the seller states that 'models are for adult collectors – not suitable for children'; Figure 6.5).

Building on this point, however, it is instructive to consider those items earmarked for 'house clearance' – the 'job lots' of toys whose biographies are effaced, largely because the principal justification for their appearance on an online selling site is divestment. It is ironic, perhaps, that the financial value of these assemblages (for we can see them as such) lies *in* their divestment, when the cost of posting the items to their eventual purchaser is often higher than the selling price. Indeed, many items are never sold; and, of course, the casual viewer can never know what happens to these 'unwanted' items. But the broader observation here is that although qualitatively different in their nature and mediation, house clearance assemblages

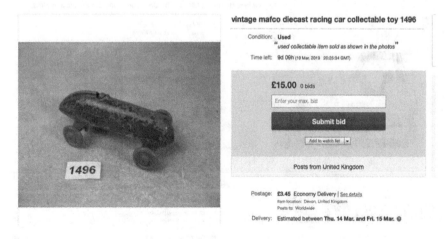

Figure 6.5 An online selling platform page for 'vintage Mafco car', clearly showing paintwork damage

on online selling platforms could, in a sense, in the massified form discussed in chapter 5, be viewed as the modern version of the 'cuts' analysed in Cambridge by Cessford (2017). At a broader stretch, they could be seen as the latest 'generation' in cultures of divestment that are a necessary corollary of cultures (and spaces) of consumption (Gregson et al., 2007; Collins, 2014). In the absence of midden heaps, wells or other spaces to create 'cuts' in many modern settlements – and in the context of the social, environmental and legal difficulties of dumping rubbish in countries like the United Kingdom – digital spaces become vital repositories for the stuff that accumulates in everyday life, including children's toys. Thus, to the intersections of manufacture, transport, popular cultures and more listed above, we can add practices and media of *divestment* to infra-generational analyses of childhood materialities as yet another stage in the lives of objects. In other words, not only do objects 'have' biographies (Crawford, 2009) but, at a macro-spatio-temporal scale, the *natures* of those biographies themselves shift, in a manner that we could readily term 'generational'.

Cuts *into* objects: on the importance of 'damage'

Looking again at Figure 6.5, one's eye is immediately drawn to the surface of the car – to the incomplete paintwork that shows the diecast metal underneath. Indeed, several items in our 'vintage' category exhibit signs of wear and tear (this also went for many dolls, such as the Zapf doll appearing in Figures 5.3 and 5.4). Not only are sellers fairly open about these defects, but many direct would-be purchasers to the pictures themselves, using a disclaimer such as '[o]ur photos are a part of the description . . . please study them carefully . . .' (part of detailed entry for item 3a in our 'vintage' collection – a 'Vintage Buddy L Bulldozer BL Corp Japan Toy Car Plastic and Metal'; Figure 6.6).

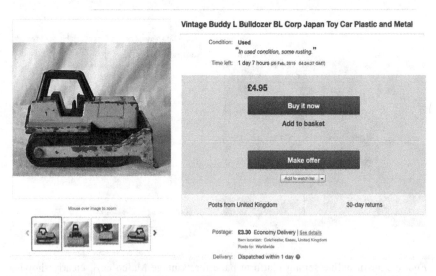

Figure 6.6 Vintage Buddy L Bulldozer, with obvious signs of use and rusting, posted on an online selling platform

Unsurprisingly, some of the older toys being sold on an online selling site have been subject to damage through various interfaces and forms of unit operations (Bryant, 2014) – whether from playing and/or the effects of the exposure of sub-surface materials (like iron) to rusting, as even apparently inert materials like metals undergo transformation (compare Bennett, 2010). This observation also enables comparison with toys found further back in the archaeological record and reminds of Crawford's (2009) striking quotation in the previous section. Along-side the various forms of wear and tear reported for an online selling site items, traces of childhoods in assemblages raise important questions about *materiality* and *objectness*. These extend – perhaps in ways that seem so physically obvious that they disprove the point I wish to make – *beneath the surface.*

Let me provide a couple of examples. In her work, Crawford (2009) argues that – beyond the microgeographies of their deposition – a second key way in which it is possible to determine that items have been left by children is to exam-ine marks, scratches and other types of apparent deformation. At Mucking, there were significant finds of materials such as bones, ceramics and metals, each perforated with literally hundreds of holes, with no obvious function. Crawford (2009: 65) signals that these are analogous with 'buzz bones' from Viking and Medieval contexts – in both latter cases, with proven associations with children's agency. Back at the Yucatán Peninsula, Álvarez (2017) draws attention to marbles made of clay and glass, which could have been used by adults for gambling. How-ever, some marbles bear scars and scuff marks in their otherwise smooth surfaces, indicating that they had been played with (and, in turn, *these* marbles were found in household lots where children very likely played, as observed earlier). And, in the United Kingdom, reflecting on 71 objects classified as toys from 1000 AD to 1600 AD in the England and Wales portable antiquities scheme, Harper (2018: 95) notes the prominence of damage to items as evidence of their toy-hood: "[w]hilst these could be accidental or post-deposition breaks, we may be seeing toys that were deliberately broken by the children who used and discarded them" – whether just for fun or to rebel against adults! In other words, alongside a range of other 'environmental' factors, pre-, during or post-deposition (including various forms of weathering and its physical location), evidence of an item's toy-hood – and its *likely* association with *childhood* – is often found in its incomplete, damaged or otherwise degraded state.

It is on this point that, although not without a little trepidation, I wonder – speculatively – about how this discussion of object biographies and infra-generations might speak back to, or articulate with, OOO. Harman (2010a) argues that Heidegger sees objects as silent, unobtrusive tools that only come into view *as objects* when they trouble us – when they break or go wrong. But, Harman argues, Heidegger's framing of objects as 'tools' is misleading because it effaces the wider category of objects, and the duality between any object's surface aura and its hidden, silent reality. As I argued in Chapter 2 – via Harman's examples of the mountain – the problem for OOO theorists is one of how to think both the status of objects and the relations between objects (whether human or not) if their interior is always somehow hidden or withdrawn. As Gratton (2014: 87) puts it: "objects as Harman will call them, have a sensuous relation, while also having qualities in themselves hidden beneath this or that side of itself" (p. 87). It is here that OOO diverges from Latour and, indeed, from most brands of new materialism, because "things must be partially separated from their mutual articulations" (Harman, 2010b: 75).

Building on the arguments around the doll analysed in the previous chapter, the question I want to pose, however, is: what (if any) challenge to Harman's conception of objects' withdrawn qualities is embodied in those objects that have been *broken, cut-into, weathered, worn* or *damaged*? In other words, how should we conceive those objects that, subject to the exigencies of *infra-generational* processes, do not resemble the glossy, spherical boundedness of a billiard ball? Clearly, there is a tension here: it is tempting on the one hand to see the kinds of damage reported above as a glimpse into the otherwise-normally obscure interior of an object; on the other hand, it could simply be argued that such 'cuts' represent the new 'surface' of any object (just as a landslide on a mountain face might, in some ways, not alter Harman's conception of the mountain's withdrawn qualities). However, I do not think that either OOO or any of the new materialist literatures more commonly deployed by childhood scholars quite have the language to deal with the various forms of *damage* undergone by objects, whether bestowed solely by children and/or a range of taphonomic processes.

Ultimately, then, I am perhaps making a straightforwardly empirical argument about the value of thinking and working infra-generationally. That is, however one might theorise damaged objects, it is through a (pull) *focus* on such objects, through an attentiveness to the very material nature of that damage and what could have caused it, that new narratives can be generated. As I have already demonstrated, these narratives require a kind of wilful ignorance of children – if only to enable perhaps radical, perhaps speculative, but certainly substantial accounts of past childhoods to emerge. Critically, in moving beyond some of the predominant frames of childhood studies scholarship, these accounts enable a storying of the after-lives of children that weaves together multiply timed and scaled processes. Moreover, these forms of storying disclose profoundly important (if hitherto hidden) ways in which childhoods are co-implicated in social, economic and/or technological changes and *trauma* (with the case of Shabona Grove being a guiding example).

Finding children III: after-lives of bones

The previous section illustrated how various forms of damage might be points of articulation for the multiple spatialities and chronologies of infra-generations. As I argued, to think in this way might enable an expanded purview of how we might think and do 'after' childhood – both in terms of how children's role in the biography (or generation) of objects might come *after* adults' and in terms of the after-lives of objects divested by children who are no longer children and, in most cases, are no longer alive. This latter observation raises a further challenge – and it is a challenge, both conceptually and ethically – when it comes to the after-lives of childhoods. Specifically, this is because archaeologists are not only interested in the material cultures of childhoods but in the material remains *of* children – in children's skeletal remains. In this final section, I want to ask how a focus on the bodily traces of past childhoods, found at archaeological sites, might push even further the notion of infra-generations. I also want to pay some attention to the objectness of skeletal remains as they are subject to taphonomic processes and to their implications for thinking about forms of *human* trauma and finitude.

A starting point – as with many archaeologies of childhood – is with burial practices and the place and status of children's skeletal remains. Bengtson and O'Gorman's (2016) analysis of the Oneota group, who lived and moved within the Central Illinois River Valley of what is now the United States during the fourteenth century AD, is particularly instructive. It is a useful touchstone because it starts with a familiar premise for scholars of more contemporary childhoods: the development of a child-centred approach that, in particular, stresses children's agency. In this case, Bengtson and O'Gorman focus as much on the objects buried with children as children's skeletal remains. But, in doing so, they highlight some significant premises for further extending infra-generational analyses. Critically, Oneota burial practices offer an unprecedented window into forms of 'cultural fusion' that were the result of the continuous and complex migratory practices of the group (Bengston and O'Gorman, 2016: 23).

Objects buried with children, including pottery, indicate a unique blending of Oneota material cultures with those of another group – the Middle Mississippian group. These artefacts are fascinating in themselves for their decoration – not least the combination of hand and bird symbols from the two groups. Yet they are also significant for what they tell us about the multiply timed and spaced, infra-generational, migratory practices of the Oneota and their children (see Georgiadis, 2011, for a contrasting example from Ancient Greece). The Oneota moved between ecologically rich environments in the River Valley, but their migrations were patterned by environmental conditions – by seasonal as well as longer-term cycles of growth, resource availability and depletion, weather, flooding and more. These movements were also served by the climatic conditions of the period (a point to which I shall return below). Entangled with/in these 'environmental' chronologies were the decisions of the Oneota to move and their inter-relationships with other groups such as the Middle Mississippian – which were sometimes peaceful but sometimes violent. Whatever the tenor of these relations, Bengtson and O'Gorman highlight that the cultural fusion of these two groups was not just evident in their material cultures but in inter-racial mixing between the two groups. This 'post-migration ethnogenesis' (Bengston and O'Gorman, 2016: 26) is significant because it demonstrates multi-racial histories of the United States that predate colonialism and subsequent (and well-documented) concerns surrounding race that continue today. In fact, it is significant for two reasons: firstly, because it highlights how burial practices surrounding children are entangled in and constitutive of a range of infra-generational processes that weave together the human and non-human, as well as the social-cultural and the 'bio' of the human; secondly, because attentiveness to such infra-generational processes might (again) resonate with critical theorisations of race that are a source of inspiration for this book. Indeed, I turn again to weavings of longer-term (infra-)generations with questions of race in the next two chapters.

The Oneota example offers some speculation (mainly on my part) about the relationship between even longer-term genetic and environmental-climatic conditions, childhoods and human generations. To what extent, though, might it be possible to 'claim' (for instance) climatic variability as a *part* of infra-generations that combine the human and non-human? It is likely 'unscientific' to claim that climatic cycles and changes are 'generations' yet they share the kinds of 'phasing' of human generations that both enters into but also exceeds human existence (see discussion of Morton's hyperobjects in Chapter 4).

In response, a look at bioarchaeological studies of children's skeletal remains is compelling for what such studies reveal about the co-implication of human generations with more-than-human processes at a wide range of temporal and spatial scales. In the first of two instructive examples, Fulminante (2015) examines the duration and eventual cessation of infant breastfeeding in Middle Ages Mediterranean countries (also Jay, 2009). This example is significant because it resonates not only with a recent turn to examine the geographies of infants (Holt, 2013, 2017; de Campos Tebet and Abramowicz, 2018) but with some of the earliest work in especially feminist-inspired geographical research on children, which focused on mothering practices and cultures (Holloway, 1998). Yet, in both scale and method, Fulminante's work also differs in some obvious ways. Fulminante combines bioarchaeological analyses of children's skeletal remains (specifically the presence of nitrogen, oxygen and strontium) with ancient historical texts to interrogate the widely held view that breastfeeding duration declined with increasing levels of urbanisation. This is significant from a feminist perspective because this assumption has always been based on (elite) men's contemporary prescriptions for those women, found in medical texts that inevitably – but problematically – predominate as surviving evidence for breastfeeding practices.

What is also significant and compelling about Fulminante's analysis is how this ostensibly political argument is entangled with a consideration of large-scale geographical and historical differences. In other words, she offers a fascinating sense of how levels of skeletal nitrogen, oxygen and calcium in children articulate with cultural and environmental changes, in different countries, across different historical epochs. There is some evidence to support earlier assumptions about the role of urbanisation in curtailing breastfeeding, but the picture is far more complex. For instance, during 'Prehistory', in Turkey, weaning started at one and was complete by two, during a period of early urbanisation; in Poland, weaning started at 6 months (i.e. earlier) but was completed by 3 years (i.e. later) and was thus prolonged in what was a pre-urban society. During the Neolithic era, regardless of geographical region, the cessation of breastfeeding took place at around 2–3 years. At the height of the Roman Empire (on which more in a moment), there were differences between the core and the periphery. These came down *partly* to the prescriptions of different physicians (i.e. regardless of levels of urbanisation) but *also* to socio-economic traditions as these continued, changed or ceased across multiple generations. Thus, in Britain, in the Roman periphery, breastfeeding duration decreased with levels of urbanisation, and especially with the move from hunter-gatherer, to agricultural, to urban societies. As Fulminante

(2015: 42) puts it: "agricultural and hunter-gatherer populations tend to practice longer breastfeeding times compared to urban populations to increase birth-spacing and control population size". Whilst the role of urbanisation is not necessarily disputed in terms of general trends, this, then, is a picture of far greater complexity than the prescriptions of elite male physicians would imply. It is also, though, a striking example of the *long durée* of breastfeeding practices in which – obviously – children are implicated. Indeed, although one might justifiably baulk at the rather dispassionate view of the intimacies of breastfeeding as not merely central to the survival of successive human generations but to shifts between different 'population' types, the term rather provocatively situates infra-generational analyses within far wider spatial and temporal frames than the equally valid but smaller-scaled analyses of infant geographies being undertaken in the present.

The example of the Roman Empire, and of the 'rise and fall' of cultures – of their *generation* – is one that brings me, in a rather oblique way, back to the questions of climate change with which I began this chapter. This is, and is not, a happy accident: on the one hand because it absolutely highlights the pressing spatialities and temporalities of climate change as hyperobject, which I discussed via Morton's work in Chapter 4; but, on the other hand because it references climatic variability anterior to nearly two millennia of human impact on the earth, and particularly the acceleration of warming during the twentieth century.

Gowland and Redfern's (2010) study compares child-rearing practices in London and Rome during the latter years and eventual collapse of the Roman Empire. The first study to make this kind of comparison, it weaves questions of childhood and intergenerational care together with processes of socio-politico-cultural change, climate change and, just as significantly, children's health. The study focuses on skeletal markers of health in children's remains found in London and Rome, namely *cribra orbitalia* (porotic lesions of the skull that lead to spongy/porous bone tissue, which indicate iron deficiency and other markers of malnutrition) and *enamel hypoplastic defects* (caused by poor nutrition alongside a range of other factors). In parallel with developments in social studies of childhood that emphasise the 'biosocial' (Ryan, 2012; see Chapter 1), Gowland and Redfern (2010: 17 & 18) offer a "biocultural approach to skeletal remains" that demonstrates how "[t]he skeleton as part of a once-living individual becomes, to an extent, imprinted by the lived social and physical environment in terms of both the chemical composition of the bones and their macroscopic appearance". Moreover, and to repeat an argument made above, children's remains not only help us to understand their own health but also that of the wider population – not least because these markers are retained in adult skeletons.

As a result, when read against the Brazilian and Danish case studies with which I opened this book, Gowland and Redfern's analyses offer insight into the *intractability* of the challenges that are and (in this case) have been faced by children. But, crucially, they also exemplify some of what I think it takes to think and do, after childhood, particularly through infra-generational styles of working. Their work is striking for the systematic but also eclectic way in which a range of processes are entwined. A critical consideration and, again, a familiar starting point for scholars of childhood and age is a series of observations about the positioning (or 'social construction') of gendered and classed childhoods within Roman conceptions of the lifecourse. In Roman times, childhood lasted to puberty (12 in girls, 14 in boys); adolescence for girls lasted as they underwent preparation for marriage into twenties; for boys, *Juventus* (youth), it could last into their twenties, depending on their social status. Thus, thinking of the socio-cultural drivers of infra-generationality, "as a society's life course is culturally grounded, it is subject to temporal and spatial differences, which are attested in the Roman Empire" (Gowland and Redfern, 2010: 19).

Gowland and Redfearn's analysis then moves to the rather different conditions found in London and Rome during the 400-year period in question. Again, this extends beyond the ordinary purview of most analyses of human generations and, in its attentiveness to the conditions of life and (ill) health in the two cities, necessarily witnesses the more-than-human materialities and processes that constituted life itself. During this period, London was experiencing rapid urban expansion as the Romans colonised Britain. Most buildings were constructed of timber, with houses sheltering mice and rats and yards for chicken, pigs and the storage of rubbish. *Enamel hypoplasias* were lower than in Rome, although not significantly so. At the same time, Rome – as a more established city – had become very populous, with awful hygiene, overcrowding and high levels of infectious diseases. The population was disproportionately youthful, because poor levels of health meant low life expectancy (between 25 and 30 years), but the population was sustained by very high levels of immigration (which in turn compounded disease-related issues with the introduction of new pathogens). Thus, consistent with these conditions, rates of *enamel hypoplasia* and *cribra orbitalia* were higher than both London's and the rest of Italy's, although the latter could also be ascribed to the prevalence of malaria in Mediterranean countries.

Critically, and putting Gowland and Redfean's (2010) study into even broader context, these health issues also intersected with climatic variability and, in turn, with a range of political and fiscal conditions that created an intractable and, ultimately, insurmountable challenge to the Roman Empire. It has been suggested that the advance of the Roman Empire – especially into Britain – took place during a period of wet and warm summers that "influenced agricultural productivity, health risk, and conflict level of preindustrial societies" in positive ways (Büntgen et al., 2011: 578). However, a period of greater climate variability in the final 400 years of the Roman Empire is likely to have contributed to its demise (Büntgen et al., 2011). Significantly, *in tandem* with the various processes outlined by Gowland and Redfern (2010), alongside ensuing political instabilities, climate change contributed to the prevalence of the very diseases that led to the markers of *enamel hypoplasia* and *cribra orbitalia* found in children's skeletal remains (Harper, 2017). This observation sits alongside more contemporary analyses of the links between climate change and human health, which range from exposure to extreme temperatures, to vectorborne and water-related diseases, to food safety and mental health (EPA, 2019) – and to which children and youth are particularly susceptible (UNICEF, 2014).

Conclusions: infra-generations, taphonomic childhoods and the offspring of alien phenomenology + speculative fabulation

This chapter began with some pressing *concerns* about climate change – conceptual, ethical and political. It ends not so much with an answer or even *a* response but a set of responses, or openings, that connect with other threads in this book, and beyond. Focusing for a moment on the #climatestrikes and the generational politics and rhetorics surrounding climate change, I want to be clear that, in thinking and doing, after childhood, via the notion of infra-generationality, I am not seeking to move away from or evade questions of climate change or of humans' response-ability for it (also Bennett, 2010). Indeed, my foray through a range of archaeological literatures ended very deliberately with the example of ancient Rome and its demise. For, although climate variability during the final 400 years of the Empire may not, or may only barely, be ascribed to *anthropogenic* climate change, the point has been forcefully made that the *effects* of that change were so profound because Romans – and especially their political leaders – ignored climatic changes and failed to mitigate for them (Büntgen et al., 2011). For many activists, scholars and commentators – including the millions of young people taking part in climate strikes around the world as I wrote this book – this is a familiar and terrifying refrain. So, amongst discussions of multiply timed and spaced processes of intractability, and of complexity, I want to argue very simply that, for the range of reasons that I and others have outlined, conventional social-scientific notions of 'generations' are insufficient in providing the conceptual and analytical tools necessary to grapple with climate change and other forms of what Morton (2013) terms 'temporal undulations'. There is, I think, a series of gaps between the discursive-affective deployment of generations in climate change debate (which in its oscillation between extreme hope and fear can stultify more diverse forms of action), the analytical properties of most generation-thinking (which are too hemmed in by their human-centredness and modest spatio-temporal scales), and the scales and registers in which climate change phases into human life as a hyperobject (Morton, 2013).

Thus, building on a set of concerns that I raised at the end of Chapter 4, I have demonstrated in this chapter how the concept of *infra-generations* may offer significant analytical purchase for decentring not only children, not only humans, but the phenomenological horizon of traditional social-scientific methods for studying generations. I have argued that infra-generations combine multiple, overlapping spatio-temporal scales, folding and co(i)mplicating human/non-human processes as diverse as urbanisation, diet, ethnogenetics, climate change, seasonal resource availability, object biographies and the temporalities of cultures (from the Roman Empire to popular cultures in the twentieth-century United States). We might not be used to thinking of all of these phenomena *as* generations; it might feel scientifically inaccurate, or just plain wrong, to do so. Yet if we can imagine – if we can speculate, even for a moment – that objects have generations (their 'toy stage', for instance), or that climate change is both generational and generative (sometimes in the most terrifying ways), then it is easier and less radical to imagine that human generations are always more-than-human: below; behind; lower; marginal; *after*. And, as I have demonstrated, infra-generational thinking can be achieved by arts of (not) noticing children (and vice versa); by the 'pull focus', in which ignoring children and/or childhoods for a while can be a necessary step to *extending* our analyses in terms of how, where and – especially in this chapter – *when* childhoods matter, whether in conditions of play, or trauma, or whatever else besides.

Conceptually and methodologically, I would argue that this chapter in some ways embodies the (queer) offspring of Bogost's (2012) 'alien phenomenology' and Haraway's (2011) 'speculative fabulation' (although perhaps the two are not so *very* different). I expand on my use of alien phenomenology in Chapter 8, and on speculative fabulation in Chapter 7, so I confine myself here to a couple of initial reflections. The first is on the status of the 'objects' represented in this chapter – and, especially, toys and children's skeletons. I have sought – as Bogost (2012) argues – to tell stories of infra-generations not (just) from children's perspectives – in terms of their 'voice' (Kraftl, 2013a) – but from the perspective of manifold objects. I would not (yet) claim that the stories told in this chapter offer phenomenological accounts of what it is like to *be* a toy car, or a marble; as Bogost (2012) argues, drawing on Von Uexküll's (2013) ethological adventures, that requires other methods still, which I experiment with in Chapter 8 in thinking- and working-with plastics and other micro- and nano-materials. Yet, if nothing else, this chapter *has* ushered in a shift in perspective, *starting-with* toys, with an online selling site and with children's skeletons. I argued that it might be possible to tentatively suggest that, in some ways, the damaged, pierced, punctured, scratched, rusted objects that have at some stage in their biographies been 'toys' challenge the smooth veneer of surface/depth in OOO. Certainly, and conversely, one might also work *with* OOO to think of children's skeletal remains as objects that have been withdrawn: literally, as what lies beneath, buried for hundreds or thousands of years. Moreover, despite archaeological tests that (again) will enact cuts or piercings into those bones in order to test them for traces of childhood illnesses, many aspects of the embodied and social lives of those children will remain withdrawn to our contemporary interpretations.

These are, then, what might be termed 'taphonomic childhoods', subject to forces of degradation and fossilisation, of literally *becoming-earth*. Morally, emotionally, we must not forget that these skeletons are bare traces of past generations of humankind: children. Yet those children are destined to remain other: not just in the sense that it is hard, if not impossible, for adults to access the worlds of children, whether through memory or careful questioning (Jones, 2008), but *also* in the sense that the material remains of those childhoods can only show so much – whether through surface accounting (of wear and tear, or the material-cultural artefacts buried with bones) or attempts to deduce the traces of illness, work, diet or lifestyle that lie beneath. Akin to Meillassoux's treatment of ancient fossils, these taphonomic childhoods are an unsettling reminder that some elements of childhood will not only remain other to but *after* our contemporary phenomenological horizons: they are beyond the sensible (and the sense-able), and especially those modes of sensing that social-scientists of childhood prefer.

Alternatively, then – or, at the same time – I have, in this chapter, sought to dig into what lies beneath by re-reading and juxtaposing new materialist and object-oriented philosophies, archaeologies of childhood and 'online' archaeologies. Because taphonomic childhoods sit generally beneath, beyond, lower and slower than the registers to which many scholars of childhood are habituated, I have sought to introduce a degree of *speculation* and imagination into the narrative in this chapter. This was inspired, in part, by the creative (but, importantly, still robust) approach of archaeologists who use, meld and narrate eclectic sources to straddle biosocial dualities in understandings of past childhoods. Perhaps it would do them a disservice to suggest that they are speculative – and perhaps it is my re-reading of their work that should be labelled such. This chapter has, however, offered an attempt to grapple with what Haraway (2011: 1) calls 'worlding': speculative fabulations that weave entangled technocultures. Yet the examples in this chapter do not heroically or sensationally announce "new worlds, proposing the novel as the solution to the old, figuring creation as radical invention and replacement, rushing towards a [geo-engineered] future that wobbles between ultimate salvation and destruction but has little truck with thick pasts or presents" (Haraway, 2011: 5). Rather, with an attentiveness to marginalised others (children, and in some cases children living with profound social, economic and/or environmental forms of trauma), I have sought to develop a kind of 'responsive attentiveness', "akin to a decolonizing ethic [. . .] accounting for generations of entangled human and nonhuman entities" (Haraway, 2011: 5). Rather than rushing towards the future in a linear teleology, Haraway argues that we may turn things around, chronologically, as per non-teleological times in Australian Aboriginal cultures: "people 'face' the past for which they bear the responsibility of ongoing care in a thick and consequential present that is also responsible to those who come behind, i.e., the next generations" (Haraway, 2011: 5). In other words, this is a disposition to care for both what lies beneath and what comes *after*. For me, care for these 'unexpected generations' (Haraway, 2011: 6) is *care for infra-generations* as part of what Braidotti (2011: 327) terms "alternative ecologies of belonging". Hereby, I propose infra-generations not only as a conceptual or methodological device for expanding the purview of generational analyses but as a set of resources (some of which have been loosely laid out in this chapter) for re-orienting ethical dispositions to childhood, generation and the earth, and to intractable challenges – like climate change – that past-present-future infra-generations face.

In the face of grand Anthropocene theorising, the speculative fabulations that I have recounted in this chapter are, really, small stories (Taylor, 2019). Even if they connect with and articulate intractable entanglements of manifold social-environmental processes, and even if they straddle huge spatial and/or temporal scales, the narratives in this chapter are not meant to be totalising. Indeed, the presence of some marbles, found as part of an assemblage, or a damaged toy car on an online selling site, encountered as part of a house clearance, or a child's skeleton, found in a burial site, pales in comparison with hyperobjects like waste (Chapter 3), concrete or plastics (Chapter 4) in terms of the material markers of any epoch denoted 'Anthropocene'. Thus, infra-generational analyses should – as I indicated at the beginning of this chapter – not be viewed as something bigger or better than generational analyses but, perhaps, as something that might invoke an even more generous, generative, inclusive attitude to the multiple spatio-temporal processes in which human lifecourses are inveigled. As such, they should be taken as a conceptual, political and ethical guide for modes of thinking and doing, after childhood.

Note

1 As outlined in Chapter 1, to conform with best ethical practice, the names of individual Twitter accounts are not provided alongside tweets. The analysis presented here is taken from a #climatestrike Twitter API scrape for the same time period as Figures 6.1–6.3 (11 February–7 March 2019).

7 Energy

In Chapter 3, I asked whether, as thinkers of relations that exceed the human (child), we are still concerned with the politics, affects and effects of matters-as-*resource*: as *resource-power*. At that point, nexus thinking offered a set of ways to attend to the multiply scaled workings of polycentric education, and young people's experiences of food–water–energy. In subsequent chapters, I have, however, become more preoccupied with matters-as-*objects* – whether as machines or hyperobjects – whose being has a certain boundedness or haecceity. Whether the ocean vortices or plastic vectors that were the objects of analysis in Chapter 4, or the toys and other rem(a)inders of childhood excavated in Chapter 6 – and whatever one's opinion of OOO – material objects were front and centre.

In this chapter, I consider phenomena that are not *counterposed* to objects but that nevertheless occupy a slightly different category of being – and which, in continuing my critical engagement with OOO in the previous chapter, again challenge the primacy of the object in thinking and doing, after childhood. I am not arguing that these phenomena are *not* material. Nor am I arguing that they are neither comprised of nor constitute material objects – whether schematised as machines, hyperobjects or as perceptible 'things'. In fact, I spend quite considerable time talking about bricks. Rather, in bringing together some of the arguments of previous chapters, I am interested in phenomena that are or could be classified via the term 'energy'. What I term 'energetic phenomena' may be material, and/or discursive, and/or constitute 'resources' of various kinds, and/or processual, and/or practiced, and/or relational. In other words, they may be all or just some of these things, in different times and places. I will be moving *beyond* but not entirely *away* from objects, with a deeper attentiveness to "social-material processes that are characteristically massy, indivisible, unseen, fluid and noxious" (Horton and Kraftl, 2018: 928). As Horton and Kraftl contend, these kinds of processes have rarely been the subject of scholarship in interdisciplinary childhood studies but, as Pacini-Ketchabaw and Clark show in their exceptional, multi-scalar, postcolonial analysis of water, are entrained within obdurate political questions (for other exceptions, see Crinall and Somerville, 2019; Rooney, 2019a; a number of chapters in Hodgins, 2019). Moreover, as I will show, energetic phenomena are also – particularly when they are combined with articulated with childhood and youth – deeply ensconced in political and ethical questions about *resource-power* and inter- (or infra-) generational justice.

Of all the resources discussed in Chapter 3 – and more – why focus on *energy*? The answer is at once pragmatic, political and philosophical. On the latter, as I will show, the notion of energy incorporates but – of course – extends beyond notions of 'resources', taking in novel theorisations of process, embodied practice, materiality, metabolism and socialisation. On the former, concerns around energy access, justice and precarity become increasingly heightened when current and future generations of children enter the debate. As the previous chapter demonstrated, the #climatestrike movement has enabled young people around the world to find a voice in relation to questions of climate change, species loss, pollution and the production and consumption of energy. Whether or not one conceives these as 'infra-generational' – and I will argue in this chapter that we should continue to – the #climatestrike movement articulates with other (inter) national concerns about the compound generational inequalities facing today's young (both *as* young people and as they grow into adulthood).

Environmental generational inequalities are also being increasingly recognised in national and international advocacy for children and young people, as well as in recent research. However, the status of *energy* is, notably, somewhat ambiguous and often rather patchy. For instance, in the United Kingdom, a recent report of the Resolution Foundation's Intergenerational Commission (2018) concludes that today's generation of youth will likely be the first to grow up *poorer* than their parents. Significantly, and symptomatic of the absence of research that connects youth and energy, the report focuses on a number of indicators around relative wealth and consumption patterns but does not mention 'energy' once. Perhaps this is because, in the United Kingdom, access to energy (in the form of electricity, gas or food) is fairly taken-for-granted – at least by the majority. Yet emerging research shows a complex, bi-directional relationship between fuel poverty and other forms of economic precarity amongst young people – both in terms of the impacts upon (for instance) thermal comfort in the home (e.g. O'Sullivan et al., 2017) and the ways in which 'energy deprivation' intersects with complex, fluid practices of energy consumption, such that 'domestic' energy poverty spills outwards beyond the home into all elements of a young person's life (Petrova, 2018).

Meanwhile, a recent UNICEF (2019) report examines the impacts of climate change upon children. Its twofold message is clear: children are the most vulnerable to climate change and related crises; and, unless we act now, the danger to children, both now and as future adults, is likely to escalate – in terms of the impacts of pollution, malnutrition and lack of access to sustainable energy sources. Indeed, unlike the Resolution Foundation piece, the report explicitly makes the link between *children* and the central significance of sustainable *energy* solutions to addressing the causes of climate change, to adaptation, and to mitigation: "[t]aking the needs of children into account in investments in sustainable energy and on climate change adaptation can benefit society, resulting in reduced child mortality, better early childhood development, improved maternal health, and better education" (UNICEF, 2019: 68). However, in its focus on investment in infrastructure for "universal access to modern energy services" (ibid.), the report takes a rather technocentric and universalising approach, where there is a need to account for the diversity of ways in which young people in different contexts experience, understand and deal with energy challenges.

In the above light, a central aim of this chapter is to divert from a focus on technocentric, engineering solutions to energy challenges, towards the telling of other, perhaps 'smaller', but certainly more diverse energy stories (compare Taylor, 2019). The focus will still – to an extent – be on climate change, although at times the links between energy and climate change might, as a result of these more diverse stories, appear tenuous. Inspired by feminist and critical race theorists, it will retain an attentiveness to the *extra-sectional* (Horton and Kraftl, 2018) processes through which energy and social difference (particularly age, gender, ethnicity and 'behaviour') co-emerge. It will also pick up two strands developed to greater and lesser extents in the previous chapter: a longer-term purview, wherein children and young people's everyday engagements with energy are embedded in longer, infra-generational energy histories; and an emphasis upon story-telling, imagination and forms of speculative fabulation (Haraway, 2011). The chapter's structure oscillates between extended empirical examples and discussion of different energy literatures. It begins with an extended series of vignettes from our Climate Action Network research with teenage boys in Birmingham, which centre upon bricks. Then, I reflect on the (dis)continuities and more-than-object concerns of those vignettes in relation to extant scholarly research about children, young people and energy. Subsequently, I move back to Birmingham, and then on to our (Re)Connect the Nexus research in Brazil, to consider more explicitly a range of energies – and energy stories – beyond energy-as-resource. In each of these cases, I explore empirical materials alongside a reading of select works that have sought to theorise energy in terms of metabolism, embodiment, socialisation and process. Doing so enables me to make an argument, in conclusion, about *energetic phenomena, after childhood.*

Balsall Heath, Birmingham, September 2018–March 2019: bricking (it)

This section is all about bricks. Literally. It is a series of stories that are loosely (if, sometimes, at all) *about* bricks. Individual bricks – in their present-at-hand object-ness, their *brick-ness* – matter relatively little. There is some discussion of the material qualities of some rather old bricks. But these matter less than the aura of brickness, and attendant practices of construction/destruction, which permeate a series of moments, encounters and *energies* across a period of months. Evoking a non-representational twisting of 'space' into 'spacing' (Malbon, 1999), 'bricks' may be enlivened by a sense of 'bricking' (a little like Rooney's [2019b] lovely evocation of 'sticking', as children experiment with sticks in the woods). Perhaps, to mash up the colloquial English for a sense of foreboding and anxiety with Haraway (2016), we are 'bricking it' – staying with the troubling capacities of bricks as much as their 'constructive' affordances. If this appears to be an odd elision of *silliness* with *trauma*, then that is quite deliberate: the juxtaposition of different forms of bricking (it) in this section is a key example of the tension between those two terms that I elaborate in the book's conclusion.

Yet, even as it is concerned with moments of speculative fabulation, this section is not concerned with imbuing bricks with a 'naïve vitalism' (Bennett, 2010: 63) that would animate bricks themselves as Kafkan automatons. That would reduce bricks to a singular relationship with children that would be reminiscent of child's play – in the most oppressively nostalgic sense. Neither, in using the term 'bricking it', is this section concerned with demonising the Black-British and South Asian-British teenage boys, with variously diagnosed 'Behavioural, Emotional and Social Differences' (BESD), who make up the vast majority of the school's students. Rather, with the provocations of critical race theorists like Huang (2017) in mind, I want to flip this narrative on its head: to recognise that these boys are, very often, the subject of societal fears; that they are, very often, the kinds of voices and subjects absent within climate change debates (including, it should be noted, many of the #climatestrikes); and that their modes of expression – with bricks being an enduring preoccupation – might evoke fear in some but, equally, foster legitimate knowledges about energy. What follows is a mixture of ethnographic writing from our fieldnotes and consequent reflections. Again, I want to stick with the many tensions between 'silliness' and 'trauma' – however defined and whoever gets to define them.

Bricks, then, were embedded in the story of our work with some of the young men at Balsall Heath. As I outlined in Chapter 1, our work at the school moves between multiple activities and sites, yet we often start and end in the classroom to frame our activities and the learning that could take place. And things move quickly. The narrative in this section is intended to evoke that speed, even if it appears 'superficial' at times. All of the events involved a particular group of between four to eight Year 9 boys (numbers varied with attendance, each week).

We are in the classroom. We are talking about coal. The importance of coal was reasserted through its role in the creation of homes and buildings in Birmingham, especially during the Industrial Revolution and the nineteenth century. We talked about where coal was used to bake bricks. This discussion sparked interest given that two students were undertaking a bricklaying course. We explored the idea that coals were used to create furnaces to melt glass and metal for use in buildings, which in turn could be used to contain or save more energy. Brian, the leader – an external speaker from the city – then proposed that we pull the blinds up in order to allow the heat and light generated from the sunlight to access our classroom. We all get up from our chairs and start feeling the classroom: windows, walls, bare bricks, pipes, cupboards, chairs, tables and other materials in the classroom.

We sit down and the conversation moves to global warming, and rising sea levels and how this will affect Britain given that it is an island with some very low-lying land. One of the students made a comment about ownership of the earth, more specifically that "the water will take back the land"; this was a comment he had made previously in a different session.

Half an hour later, we went on a guided walk around Balsall Heath, guided by Brian, a local environmental activist and community leader. We proceeded to visit the railway bridge, outside the main school site (Figure 7.1). Brian told us about the plans to build a train station for Balsall Heath; he also spoke about a historic railway station that operated between 1840 and 1941, at the height of the city's industrial phase. The texture and colour of the bricks were different to the normal red bricks so familiar throughout the area; they were black and shiny, almost as though they had been painted. The boys spent a long time running their hands over them. They asked why the bricks were different colours. We didn't really know, although the railway bridge was a fairly conventional design, replicated all over the city (and all over Britain), given its Victorian heritage. We spent some time talking about the changing modes of travel within Balsall Heath, and what those meant for energy use: roads, walking and cycle routes.

We carried on walking around Balsall Heath – past a parade of shops and food outlets selling an array of wares – and then downhill. Brian showed us a modern church with solar panels, before the students stopped outside a building which had been built as an extension of a much older stable (this was evident through the high-up opening where the horses would be fed hay). The difference in brickwork between the old and newer parts of the building was particularly fascinating to

Figure 7.1 Railway bridge near the school, Balsall Heath, Birmingham
Source: Author's photograph

the students, and especially those who were undertaking a bricklaying course, as the older building was made from handmade bricks which had pebbles concealed within them and large cement-filled gaps between each brick. They spent some time picking at the cement, rubbing it, dislodging small stones and then running their hands over the newer bricks. There seemed, in fact, to be several phases of building, and Brian told us that this was likely the oldest building in the area. The students passed their comments on the 'unsatisfactory' brickwork! Whilst this was happening, and with our backs turned away from the street to face the building, two of the boys had somehow got into an argument with an older man who had passed us and were shouting names at him as he walked away. They then turned back to the bricks.

After this the walk tailed off a little as it started to rain quite heavily. We looked at a tree planted by the community – the boys finding this more interesting once they discovered that Brian himself had planted this. We talked about the role that trees have in cities – for shade, oxygen and absorbing CO_2. One of the boys called the tree 'a gas-guzzling motherfucker'.

Eventually the wind, rain and cold hurried the students back into the warmth of the school. We had been due to visit an ecohouse, with various energy-producing and energy-saving technologies, which is again located fairly near to the school. Given the inclement weather, Brian decided to use the extra time gained from the walk being cut short to discuss what the ecohouse would have looked like from the inside. Whilst he was talking, and apparently ignoring him, the students pulled out a tray of Jenga blocks and proceeded to build a house. The house turned into a mansion of sorts with all those present in the room, students and staff alike, assisting in building the final product. The mansion had a curved round roof, intricately placed with the aid of the other students. There was a moment of awe at the final product before Arooj (my co-researcher from the University of Birmingham) was asked to 'do the honours'; the pieces of Jenga smashed everywhere, as they put it "like a tornado hit it!"

Several weeks before the day of the guided walk, we took the same group of students to ThinkTank – a science museum in Birmingham city centre that caters for school groups of all ages. The museum ran a bespoke LEGO™ Engineers session for us. The session involved an introductory presentation to different energy sources, before we had an opportunity to work in groups to make wind turbines (Figure 7.2). I worked with Sami – a 14-year-old British South-Asian boy – and one of the teaching assistants, Alyssia. Sami's work on our construction project was sporadic. At times, he was heavily invested in selecting, examining, scrutinising and discussing the different components. At others, and without warning, he would get up and walk around the room, leaving us to continue with the build, before coming back and joining in. He seemed constantly to exude a nervous energy, always moving, shifting in his chair, rarely making eye contact and making comments that did not seem (to us) to be relevant to the task.

But once we had finished building our turbine, Sami suddenly became really careful – even more nervous – walking across the room holding the wind turbine. The idea was that we would face it into a large fan, which would power the

Figure 7.2 Making wind turbines at the ThinkTank Museum, Birmingham
Source: Credit: Arooj Khan

turbine, and then a bulb that we had connected to the turbine. We faced it into the wind, connected some small bulbs, and – it worked! There were whoops, our hair and clothes flapped in the wind, and Sami (who was very thin) almost seemed to be being blown backwards by the fan.

After that, I joined another group, who were still finishing their build. Another boy (Tony) mentioned canals – could we get energy from then? We talked a bit about what the canals were for, referring to some of the old factories that once stood (and in a few cases still stand) both near the museum and in the area around the school. There was some recognition that there was lots of energy in Birmingham in the past – the canal boats "moving stuff around" as one boy put it, and lots of energy being burned by factories. We talked about how the bricks that built many of the factories were moved by canal. Then Amir asked, "did you know the *Peaky Blinders*[1] use the waterways when they kill people?" He also noted Birmingham's strategic position within the waterway canal systems, "it's good that we're in the middle, oh I didn't know that we were so connected". However, the conversation stopped almost abruptly as it started, as one of the boys started to grab for bits of the LEGO on the tables, so we couldn't explore in any depth.

Then, from behind us, a few things happened all at once – back with another group who were also testing their turbine, the fan blew so hard that one of the sails from the windmill became detached and literally flew all the way across the room (about 20 metres), demonstrating how powerful the fan was, leading to oohs and aahs from the boys (and the adults!). Meanwhile, on another table, the group discovered kinetic and embodied energy as – upon detaching a piece from their turbine that had been fitted wrongly – the piece flicked across the table and into a pupil's mouth, to uproarious laughter (he was OK).

One of the boys who had taken a strong lead in creating the structure ran to the larger fan that was in the front of the room to test the turbine: "It works!! . . . imagine how many of these you would need to light up your whole house though." Boys from the other tables began testing their wind turbines, many commenting at how cool the wind turbine was, with one boy asking if he could take his home. Another boy was so proud of the wind turbine that he asked for his picture to be taken with it. After lunch, the boys were able to explore the rest of the museum at their own leisure – many of them spending time looking at the steam- and coal-fired machinery preserved from the nineteenth and early twentieth centuries.

Towards the end of our work with this particular group, we visited the Lapworth Museum of Geology, at the University of Birmingham. The Museum houses displays with hundreds of rocks, minerals and fossils, alongside interactive exhibits and – perhaps most popular of all – a full dinosaur skeleton. We were – on the basis of our misconceptions – a little unsure as to how successful and engaging the museum would be. However, it transpired that the museum facilitated a range of encounters and discussions, inspired particularly by bricks (or, rather, rocks) and construction (or, rather, the doing of a jigsaw).

We talked about what killed the dinosaur. The boys knew it had likely been an asteroid, with 'the crust exploding', creating a mass extinction event. We linked this to the possibility that we might kill ourselves as a species if we were not careful with the earth. Anthony (leading the session) asked 'why', and the boys made connections to climate change. We also talked about pollution and Donald Trump – who they said was a racist, wants to kill everyone and is stupid. The connection they made to global politics was interesting and completed unprovoked by any of the teachers/adults. This led to a short conversation about the difference between Trump and Elon Musk. Throughout our months with them, the students frequently referred to Donald Trump – initially, apparently, as a throwaway soundbite or a joke or an expletive. But – although articulated bluntly sometimes – the regularity with which they mentioned Trump as part of our discussions about climate change was emblematic of their criticality of right-wing environmental politics.

The fossils, in their display cases and drawers, sparked off a conversation about the future of fossil fuel; a student commented that the future sources of fossil fuels could be human remains and food waste. One student didn't quite believe me when I said that fossil fuels were made by compressing animal and plant remains under layers of earth. "I don't believe it, wouldn't it all just crush?"

Some of the boys knew some of the rocks in the museum from the online game Fortnite. In particular, they knew obsidian, copper and malachite – but they thought that copper looked different in the game and that the museum had made a mistake. They questioned different realities – the digital and the museum – and where the museum had got its facts from. Discussing Fortnite (and looping for a moment back to some of the debates considered in Chapter 5) spawned a conversation with the teacher about games use by some of the boys – they spend all night playing and then come into school and sleep. Some find a corner to sleep in and spend all day there.

One of the most popular items was – to our surprise – a large jigsaw that showed an image of the world. It was popular not just as an intrinsically interesting or fun activity but for the wider questions and discussions it prompted, again entangled with climate change energies. As we constructed the jigsaw, we talked about climate change. One statistic which really resonated with them was the incredible amount of human displacement that may occur as a result of the effects of climate change – 500 million by 2050, compared to 60 million affected by war-related displacement in 2017. "That's shocking miss, that can't happen"; "Will it really be that bad?"

Jack and Arooj started a conversation about colonialism. This was sparked by the fact that he could easily name all of the Caribbean Islands. He mentioned that he was particularly interested in the Middle East, specifically Syria, Qatar and Lebanon. When the map was completed he stood over it and said, "but I don't understand, how did this little country take over the whole world?" Jack returned to this moment on the way back to school at the end of the guided walk (in the rain). He mentioned the conversation that we had at the Lapworth about colonialism whilst completing a puzzle of the world and how it resonated with him; Jack and Arooj spoke more about the racial element of climate change and then proceeded to speak about the concept of white allies.

(Dis)connection, energy and education

The kinds of experiences and encounters involved in 'bricking it' with students sit somewhat awkwardly with previous published research about children, young people and energy. To generalise greatly (and I come back to other literatures a little later, and especially those more pertinent to the Majority Global South), that work focuses on two key concerns. On the one hand, recent scholarship has investigated children and young people's everyday energy consumption and use of energy technologies, particularly within the home. This subset of work has two features: an emphasis upon the ways in which children understand and are involved in efforts to mitigate energy consumption, with the repeated observation that they often conserve more and use less energy than adults (Yamaguchi et al., 2012; Fell and Fong, 2014; Wallis et al., 2016); and an emphasis upon children's interaction with or use of 'technologies' – from relatively banal, ubiquitous domestic features such as thermostats and pumps (Horn et al., 2015; Strengers et al., 2016) to the role of more innovative technologies, such as solar lamps (Sharma et al., 2019), Information Communication Technologies (ICTs) (Haunstrup Christensen and Rommes, 2019) and gamification (Beck et al., 2019).

On the other hand, there is a fairly established body of scholarship on energy education – although much of this work is embedded in, and is a relatively minor component of, wider literatures on environmental education and citizenship (e.g. De Hoop, 2017; see also Chapter 3). A central preoccupation has been with whether and how young people 'carry' environmental messages, translating these into behaviours or influencing (adult) others – with the picture being often positive, if mixed (DeWaters and Powers, 2011; Toth et al., 2013; Aguirre-Bielschowsky et al., 2017; Merritt et al., 2019). Significantly, within far more established literatures on models and methods for environmental education more generally, recent work has emphasised the role of *media technologies* – including mainstream films and, again, gamification (Tranter and Sharpe, 2012).

However, as I argued in the introduction to this chapter, there is a need to understand the sheer *diversity* and *complexity* of experiences and roles that children and youth have in relation to energy – beyond technology, beyond education and beyond (domestic) use. The series of vignettes from Balsall Heath affords a brief sketch of what this diversity and complexity might look like, focusing upon young people whose voices are so rarely heard in debates about energy and climate change. At first glance the bricks appear as a kind of hyperobject. They phase into and out of our conversations and encounters, in different times and places. They are ubiquitous, in a way that appears both continuous and discontinuous – threaded through the railway bridges and buildings and very infrastructures of Balsall Heath (including the redbrick Victorian schoolhouse in which the school is situated), but not the only material of which the urban environment is composed (concrete, asphalt, metals, soils, plastics, etc., all have their places too). Each of these materials has its own timings, its own *(infra-)generational* qualities and, as a result, its own generative powers. Materially, the bricks are of their era, in terms of their size, colour, purpose and physical condition (contrast those at the railway bridge with those at the old stable). Even bricks don't last forever – and probably less long than plastics as remnants of the Anthropocene (Chapter 8). The bricks offer frequent conversation-starters, and reminders, and points of inspiration, for (often brief and incomplete) forays into Birmingham's industrial past, its energetic histories, as bricks were at once built here (through coal power) and were used to build the energy-intensive waterways, houses and factories that became the material manifestation of the Industrial Revolution's insatiable appetite for fossil fuels. Evoking the concept of *infra-generational relations*, bricks are emblematic of a period of history in Birmingham that at once combines and exceeds single *human* generations, but which is intimately entwined with and constitutive of them.

To follow a brief but necessary diversion: bricks are a crucial, apparently banal, but ever-present part of the industrial-capitalist-colonising-globalising logics of the Anthropocene (or whatever it is turned) that, when excavated, might make us all *brick it*. Bricks are as caught up in this contemporary predicament as materials like steel, concrete and plastics. In Birmingham, brick-making was vital to the development of factories, warehouses and the housing for workers. The sheer variety of bricks found in Birmingham attests to the multiple, often family owned (and later agglomerated) brick works that grew up in Birmingham and, especially, in peri-urban villages in the Black Country. Interestingly, and following Bryant's

(2014) reading of Mumford's (1934) work on industrialisation (Chapter 4), whilst nearby seams of coal in the Black Country literally *fuelled* Birmingham's (and, by extension, the United Kingdom's) Industrial Revolution, it was the presence of clays at sites like Yardley and Acock's Green that helped to literally *build* the sites for that revolution (AGHS, 2019). Bricks, then, were 'insistent', in Bryant's terms: before the Industrial Revolution, brick-making was a secondary occupation on farms, with clays simply left to rest on the soil surface after ploughing. Bricks would only have been made (and dried) during the summer. But – and here Mumford's work really resonates – with the advent of steam, bricks could be dried and then fired all year round, enabling their production in greater volumes. The proximity to the growing city of Birmingham ensured a constant market, whilst the developing canals provided a conduit for the transport of thousands and then millions of bricks. Brick-making was hard work: dangerous, dirty, manual labour (on which more in a moment). Brick-making was also often small-scale, undertaken by households – such that the sheer variety of bricks produced in Birmingham is both bewildering and fascinating (from a contemporary vantage point: see Figure 7.3) but also such that families received very low prices for their products (AGHS, 2019).

As we discovered on our walk around Balsall Heath, reminders of Birmingham's brick-making industry – of the embodied energies of bodies, steam, transportation and machinery – are to be found *everywhere* in the landscapes of the West Midlands. These structures – like bridges (Figure 7.1), houses, factories, dividing walls, even the 'redbrick' heart of the campus of the University where I work – are the banal, hidden-in-plain-sight structures that frame our everyday

Figure 7.3 A collection of bricks from Birmingham and the Black Country on display for contemporary visitors at the Black Country Living Museum, Dudley

Source: Author's photograph

lives. The lifetimes of the brick extend beyond those generations of labourers who created and built with them, as an infra-generational analysis might imply. Their display in a museum might, for a moment, render them visible, although their artful display as specimens exemplary of their kind might aestheticise and anaesthetise the graft of the hands that went into their making.

Beyond Birmingham, brick-making continues into the present day and is similarly beset by overlapping, intractable and what I interpret as infra-generational challenges. The wonderful but harrowing work of the Blood Bricks team (www. projectbloodbricks.org/publications; Brickell et al., 2019; Natarajan, 2019) is exemplary of these challenges. Their work focuses on entanglements of climate change and modern slavery in the brick-making industries of Cambodia that are fuelling the (urban) development boom in that country. Owing to a changing climate, previously agricultural labourers find themselves drawn into labouring in brick-kilns that are dirty, dangerous and polluted. The kilns hold individuals and families in conditions of modern slavery for decades – for generations – through debt bondage. Child labour is rife as children spend their entire childhoods (potentially their entire lives) at these kilns: "whole worlds are created in these kiln walls: children are born, raised, and remain to become parents themselves" (Brickell et al., 2019: unpaginated). Of all the striking, troubling aspects of this work that should cause us to *brick it* is the entanglement of indenture with climate change:

> Climate change is further testing workers' hope in a future free from debt bondage. Unseasonal rains halt work as kilns are largely uncovered with rainwater risking the quality of drying bricks. In this period, workers are not allowed to leave kilns and seek other work, instead they are forced to borrow more from kiln owners to cover their daily expenses. Darkening skies means the raining down of further debt.
>
> (Brickell et al., 2019: unpaginated)

Bricks are a point of articulated for infra-generational relations of labour, climate change, urbanisation and human generations, as, in very different conditions of marginalisation from those experiences by the young men in Balsall Heath, children becoming 'geological agents' (Hadfield-Hill and Zara, 2019; also Gallagher, 2019).

Thus, *bricking it* is about more than the bricks themselves, as I have argued: it is about how bricks constitute and articulate with diverse *energetic phenomena*. Bricks are a gateway to other kinds of earthy or rocky concerns – to the fossils in fossil fuels, and the deep-geological timescales that resolve as one reaches out to touch a carefully curated specimen; to the more recent embodied energies of labour from the Industrial Revolution that one may similarly reach out to touch at a former stable in Balsall Heath, or a local museum; to the minerals that appear (however accurately) in the game Fortnite; to the injustices writ by climate change, and the politics of the Trump era; and, hence, to rather different forms of *resource-power* than those outlined in Chapter 3. Bricking it was also a little more metaphorical: a cipher for the different kinds of construction (and destruction) in which the boys engaged. The bricks were redolent of their interest *in* construction, which became a key 'hook' for us for our varied, stop-start conversations about climate change and energy. Construction and destruction were also key activities in themselves – it was the act of *building* the LEGO™ turbines and the world jigsaw that was so valuable, and of building and *destroying* the Jenga house that was so telling.

On the latter, the narrative above was not quite 'complete' when it came to the Jenga house, and especially the reference to the tornado. This *could*, in many conventional interpretations, have been put down to a random act of destruction by 'disruptive' boys. However (or, perhaps, in addition), the previous week, the boys had visited Brian's house, which, along with a number of other properties in Balsall Heath, was actually hit by a tornado in 2005. Land-based tornados (as opposed to water spouts over the sea) are fairly rare in the United Kingdom and seldom cause the kinds of damage that they do in other contexts. Nonetheless, this tornado has been linked to both climate change (and an increase in extreme weather events) and urbanisation (because the unique 'heat island' effect in Birmingham may have contributed to the energy required for the tornado to become so powerful).

In relation to environmental and energy education literatures, I could make a series of fairly obvious observations about the value of kinaesthetic learning, and the (possible) trade-offs involved in using LEGO™ – a globally branded, *plastic*-containing product, but I suspect that simply calling these out here will suffice. Instead, I am more interested in what the structure and rhythm of the above narrative – which speaks, to an extent, to the structure and rhythm of our months with this group – can tell us about how these boys engaged with energies of, and beyond, climate change and resource-power. As I have indicated, the narrative speaks of a kind of phasing of bricks, of construction and of destruction, in and out of focus. But it also attests to the kinds of learning, the kinds of encounters and the kinds of conversations that emerged, in an often non-linear, non-teleological way. There were connections, backwards and forwards in time and across different temporal epochs. There were also disconnections, disruptions, moments of *silliness*, uproarious laughter, frequent expletive utterances, stalled conversations and multiple loose ends. Students walked off in the middle of a task. Others literally fell asleep, thanks to many nights playing Fortnite until the early hours. Multiple energies – embodied, affective, material – were at stake in these encounters, and in the next section of the chapter I focus on this observation in much greater depth.

But, before continuing, I return to the point about intersectionality (or extrasectionality), acknowledging the school's mission to work with teenage boys excluded from mainstream schools, largely as a result of 'BESD' deemed too problematic or challenging for mainstream contexts. What might generally (in the UK mainstream educational sector, at least) be understood to be forms of 'silliness' and 'challenging behaviour' were woven in/through learning encounters. I hesitate to celebrate or romanticise these, but nor do I want to efface them; rather,

I want to question *what place* such behaviours, dispositions and forms of youthful agency might have in questioning or propelling different forms of climate change politics (and, indeed, why these might be seen as less or more acceptable than acts of 'silliness' and play evident amongst younger children at the Trust's playscheme, or amongst activists taking part in climate change demonstrations).

As Rosen (2015: 39, original emphases) writes of the 'scream' in the very different environment of the nursery,

> screams [are] part of the 'soundscape' . . . they are overflowing with meanings including about inequities in the social order of educational settings. These meanings are afforded by the physical and sociocultural aspects of voice quality, as well as *over*civilizing efforts. Suggesting an approach of methodological answerability in listening to 'the scream', [we might consider] voice quality in relation to what matters and as a *mode* of potential transgressive and political articulation.

How might we become answerable to the sudden utterance of a black, 14-year-old-boy, excluded from mainstream school for his behaviour, when he shouts that a tree planted to absorb CO_2 is a 'gas-guzzling mother-fucker'? And how might we become answerable not only to Greta Thunberg (who, notably, talks of struggles with mental health) and the (in many cases) whiteness of the #climatestrikes and the environmental movement more generally but to the extra-sectional energies of race, generational positioning, emotional/behavioural/social difference and the materialities/practices of 'bricking it' in Balsall Heath? With their knowing critiques of British colonialism in mind, how might those boys' *voices* matter in ways that go beyond conventional understandings of voice (Kraftl, 2013a)? I do not have any ready answers – these are open, perhaps intractable questions, after all – but in an attempt to broach them I seek to complicate the picture further by telling some more stories of diverse energetic phenomena.

Diversifying and complexifying energies: from Balsall Heath to Brazil

Alongside our work at the school, we also worked with a nursery and playscheme run by the St Paul's Trust in Balsall Heath. Together, the two sites catered for children aged between 2 and 10, who came overwhelmingly from the more local, British South-Asian community. As at the school, we engaged children in a range of activities designed to explore different forms of energy; as the observations and reflections below demonstrate, most were loosely arts-based, although we also worked together on a range of making/doing/technology projects. Consequently, the main aim of this section – which reflects both on our work in Balsall Heath and some of the visual webs from the (Re)Connect the Nexus project in Brazil – is to complicate and diversify the ways in which children, young people and energy are (or could be) co-implicated. I juxtapose these two projects because both entailed a commitment to creative, collaborative and *visual* approaches to thinking-through *energetic* phenomena *with* children and young people. Again, the narrative takes the form of a loose series of vignettes before some deeper conceptual reflection.

Balsall Heath

One of the first activities at the nursery involved a large sheet of white paper, set out in a low-sided tray. On the paper, we squeezed out blobs of coloured paint. These were covered with cotton wool pads. The children then worked in small groups to hit the pads, in order to explore embodied energy (Figure 7.4). We also introduced a related task with elastic bands (to explore elastic potential energy), paint and large pieces of paper. The outcomes of these tasks were both very messy and very colourful, with the splatter pictures being displayed on the nursery walls.

Each group took the task in different directions, playing different games and using the hammers in different ways. In one, the children developed their own game of lifting up the cotton wool pads to say 'wooooohhhh' at the hidden colour. Negotiations were made regarding ownership of one hammer in particular, as there were only two plastic hammers amongst three children. The children then started swirling the paint together. One boy continued his findings of the more energy the bigger the impact by repeatedly hitting the cotton wool pads, resulting in larger splatter patterns than the previous group. It almost turned into a game

Figure 7.4 Exploring embodied energy through paint splatters at the St Paul's Nursery, Balsall Heath

Source: Arooj Khan

between the brother and sister: who could hit the most cotton wool pads the hardest!? The cotton wool pads melded into the paper to create a 3D abstract paint piece of sorts. The force exerted resulted in paint splatters on the workers' and children's hair, faces and clothes, as well as the grass and up the sides of the tubs that were being used.

Doing the task with the older children at the playscheme, we asked the group if they had heard of the word 'collaboratory' – as expected, they hadn't (see Chapter 1 for a description of this term and the methods for the Climate Action Network project). We broke this down and asked if they had heard of the word 'collaboration' – again, they hadn't. We then asked about the word 'laboratory'. A few of the children stated that they had heard of the word. "I have a science lab at home!! I have goggles, a coat and I can make slime and look at chemicals!" The word 'collaboration' was of particular interest; it triggered a kind of word association rhyme/song that was repeatedly sung for the duration of the session "collaboration, compilation, commiseration".

So, we pose the question here: was this activity a 'success' or a 'failure' when it came to learning about different forms of energy? In particular, what to make of the game-playing, the mess and the song?

A later activity at the playscheme involved learning about renewable energy sources, but also similar reflections on 'success' and 'failure'. We attempted to build solar-powered lamps with the children. Explaining solar power was difficult. This resulted in the use of a story format to explain the earth's dependence on depleting non-renewable energy sources and the need to utilise renewable energy sources instead. "I know this!" said one child, "my mum says that it's good to recycle too because it helps the earth otherwise we have too much rubbish and the animals eat it and die". Another intervened: "I'm excited for this! We have solar lights in our garden too, I'm going to put my jar in the garden with the other lights I think. . . ."

At the end of the session, those who took part in the activity stood outside the farm house in their eagerness to collect their lamps. There were some moments of confusion given that the lamps were not glowing in the daylight: "I don't understand, why won't it light up during the day? It should be charged up now?" Arooj demonstrated how the lights would glow by taking them into a darkened corner of the farm room. Another boy refused to leave. "I'm not going to leave! I'll sit here and watch please." His determination to stay was endearing; he became the designated helper for the activity, putting the solar cells in the garden (or "planting the lights" as he described it); bringing the lights back in; setting the table and helping us clean up at the end of the activity.

Although I talked at the end of the previous section about the non-teleological nature of some of the learning at the school, at the nursery and playscheme we introduced a task that was in some sense intended to be a culmination of the children's learning about and exploration of different kinds of energy. The task was called 'energy superheroes'. The children were asked to invent – through drawing, modelling, dressing up, acting – the superhero that they thought could deal with the world's energy problems. Each superhero should have energy superpowers and generate energy in some way. Some of the examples can be seen in Figure 7.5.

Figure 7.5 Energy superheroes at the playscheme at Balsall Heath
Source: Arooj Khan

Most of the children resolved to make masks and then to act out their superheroes' powers. The choices of superheroes were influenced considerably by popular culture. In just one example, one girl chose her mask and settled down; however, she soon grew despondent and said, "but I like Doctor Who, I watch him every night but Daleks don't wear masks!" She left the table and then eventually came back to continue with her mask, having seen others complete theirs. "Are Daleks your favourite?" Yes!! She began to shout out "Exterminate!" repeatedly and loudly before being told to be quiet by another child. "Why do you like Daleks so much?" "It's because they have a death ray arm and they look so funny!" "They have a death ray arm?! How do they get the energy to use a deathray?" She looked at Arooj as if it were obvious: "It's because they put themselves on charge." She then proceeded to run over to the dressing-up box, before asking for Arooj's help with the Velcro at the back of a dalek outfit; she then placed the mask over her eyes and ran back out towards the pitch shouting "Exterminate!" and holding her arm out in a death ray fashion.

Later, once they had finished their masks, the children led Arooj further up the path and into the covered area of the 'headquarters' (the covered area visible in Figure 7.5). The girl pointed to the bench and said, "So miss, this is where we sit and talk about how we fight bad things. Over here are the screens, it's like CCTV we can see everything." "Yeah and here are our gadgets and our batmobile."

"Miss, there is a lot of electricity in this headquarters but we're using it for good things!" They promptly acted out a scene within which the boy spotted a villain on the 'CCTV'; they proceeded to pull their masks down, pick up some gadgets and ran out to the pitch to fight crime, remaining eco-friendly by leaving the bat mobile behind.

Isolating energy as and beyond 'resource-power' in Brazil

Before reflecting on the Balsall Heath vignettes in greater depth, I want – with the discussion in Chapter 3 in mind – to turn fairly briefly to some of the ways in which young participants in the (Re)Connect the Nexus project articulated energetic phenomena. As in Chapter 3, and as a comparison with the creative-artistic activities we deployed in Balsall Heath, I present two examples from the visual web exercise that relate rather different experiences of energy – especially in terms of the urban (in Adriana's case) and rural (in Vitória's case) contexts in which the two young people live.

Adriana is a 21-year-old female from Guaratinguetá (the town that figured heavily in Chapter 3). She describes herself as lower-middle class. She is a 'worker' – a process analyst in a company based near her hometown. Adriana combines work with studies in a technology college in the evening, where she is attending the *cursinho*, a preparatory course for university. In her web (Figure 7.6), Adriana is particularly preoccupied with the factory in which she works. She chooses to focus

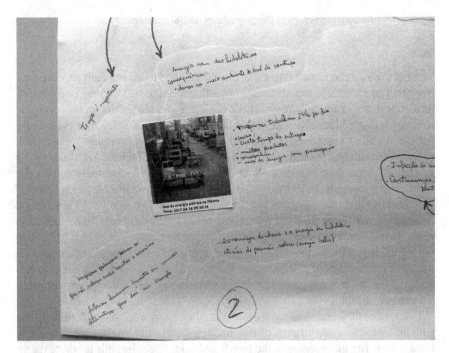

Figure 7.6 Part of Adriana's visual web: Use of Electric Energy in the factory

upon a photograph taken through the mobile (cell) phone app, adding the caption: 'Use of Electric Energy in the Factory'; she wants to show how much energy is wasted inadvertently in the factory where she works.

Adriana explains that the machines visible in the image – lathes and mills – are switched on 24 hours to maximise production. She suggested that a way of saving energy and money would be to use solar energy: producing and storing solar energy through solar panels, she suggests, would reduce consumption of hydroelectric energy, which, she says, is currently the main form of energy production and is quite expensive; this would also benefit the environment as it would avoid destroying the environment to build hydroelectric plants. She suggests that producing solar panels at a large scale would make them more affordable, and this would incentivise people to use these resources. She also notes that as factories consume larger quantities of energy, they should be the first to turn to solar energy; they should invest more in solar energy and 'give a good example'. She writes all these ideas (in Portuguese) in yellow boxes around the photo, linking them to the picture through arrows.

Alongside the image of the factory, Adriana talks about another photo – with the caption 'Brazilian Lunch in the Factory' – which is not shown in Figure 7.5. Given our nexus approach, we asked her to elaborate on the connection between food and the factory. She says that she doesn't have the opportunity to go somewhere else for lunch – which she would do if she had a car (she hasn't got the driving license yet), although that would mean spending more money, she adds. The quality of the factory food is not great. In the green textbox attached to the picture she summarises: 'greasy and poorly prepared foods; some people avoid [having food at the factory] and eat somewhere else; I have lunch in the factory because I don't have the conditions to go to another place'. Like a number of other participants in the project – notably, middle-class students with hectic university schedules, she then draws two green arrow lines linking the picture (and related ideas) to time 'Because I don't have time to eat' she explains, and to energy consumption 'for the energy used in food preparation'. Sometimes, she explains – connecting back with the students at Balsall Heath School lacking energy because they have not slept – she 'lacks energy'.

Following this flow (food/time connection), Adriana was prompted to elaborate on the photo picturing 'Afternoon Snack on the Way to School', showing the participant's hand grabbing a wafer biscuit as she goes from work straight to evening school (again not shown in Figure 7.5). Around the picture, circled in green, she writes: (right-hand side of the picture) 'I chose to eat during my way to the school as time is short and if I ate before going to school I'd be late' (bottom right corner of the picture): 'I feed myself while I walk (walking)' (left-hand side of the image): 'The transportation between the factory and the school is time consuming and with this I have less time to eat.' She says that it would help if transport were faster; for example, if she had a car she could get directly to places, avoiding the waiting/walking time, which would help saving time and ultimately make her routines more balanced, and improve her food practices too. As transport connects with energy, Adriana connects food, transport, time and the factory in her nexus of resource-power that, in contrast to many conventional analyses of the water–energy–food nexus (Chapter 3), starts with *energy* and not water.

Vitória is a 17-year-old female, also from Guaratinguetá (or, more properly, from the town's rural hinterland). She is from a middle-income, rural family. Vitória is a student in a technical school. The family (especially the grandparents) run a farm; Vitória is involved in the management and day-to-day work in the farm, particularly work with the cows and milk production. Vitória placed some images around energy centrally within her visual web (detail in Figure 7.7).

Vitória was prompted to elaborate on energy in her everyday experience starting from something she mentioned in the previous interview about using the wood-burning stove that they have at home. It turns out that her family has two main sources of energy for use in the home: a wood-burning stove and a gas stove. The wood stove serves the multiple purpose of cooking food and heating up the house; it is also preferred because it is cheaper as the wood used as combustible is available in the countryside, whereas with the gas stove the gas bottle has to be bought. The wood, however, sometimes runs out, so they also use the gas stove. Vitória also says that a downside of the wood stove is that it produces soot, so in this respect the gas stove is more practical and doesn't pollute the house. In general, however, the wood stove is preferred (mainly because it's cheaper); she also says that the food cooked on the wood stove is different (we assumed that she means that it tastes much better).

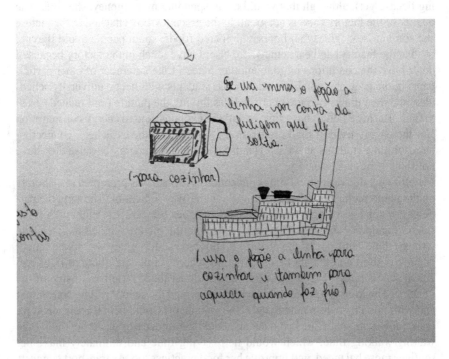

Figure 7.7 Vitória's visual web, focusing on energy and the cookstove

Theorising and juxtaposing diverse energetic phenomena

For a moment, I want to hold all of the kinds of energy in this chapter together, and in tension. This is not an objective, even a robust endeavour. I know that these stories involve diverse, in some cases unconnected children and young people, in different times and places. They are not all *related*, although they are all *characterised* by sets of multiplicitous energetic properties. For a moment, then, and following Latour (2005), I want to list these manifold properties alongside one another and point out some possible points of connection. So, in this chapter, the following 'energies' have come to the fore (amongst others, and in no particular order) . . .

> . . . embodied energy in bricks and their production, their critical role in Birmingham's collective energy histories, and the acquisition of skills to make and construct with bricks. . . .
> . . . and embodied energy in rocks and minerals, like coal. . . .
> . . . and wind and solar power technologies (of the most modest kind) . . . .
> . . . and modes of transportation, including trains and canals. . . .
> . . . and the extreme power of tornados and the intensifying energies of climate change. . . .
> . . . and embodied human energies (and lack thereof) – lack of sleep, 'violent' acts and utterances, messing about, play, migration. . . .
> . . . and embodied experiences of 'resource-power' in the 'nexus': where the timings of factory life pattern the consumption of energy and food (and the lack of time to eat). . . .
> . . . and the embodied (sensory) effects and affects of burning fossil fuels – from exposure to air pollution to the taste of smoked food. . . .
> . . . and the emotional and affective energies that produce and flow from embodied energies – anger, silliness, humour, frustration. . . .
> . . . and energy education . . . and the skills required to make and construct bricks. . . .

The list begins and ends with bricks. This demonstrates that, far from a teleology, there are circularities involved in energetic phenomena (as the basic principles of physics tells us) that extend beyond the forms of circulation considered in Chapter 5. There are also multiplicitous interconnections between energetic phenomena – some of which are also evident in the list above. There is no certain cause-and-effect connecting all of these forms of energy. Yet, in laying them out alongside one another, two things are evident: first, that the sheer multiplicity of ways in which children and energy may be related easily exceeds the often rather singular, functional and technocratic ways in which previous research and practice has characterised that relationship; second, that we need to develop new ways of theorising children *and* energy, which might enrich our theorisations of both terms – but especially, in focusing on the multiply conjunctional nature of this relationship (different variations of the 'and'), that might set out further ways of thinking and doing, after childhood. I want to end this chapter by focusing on three (by no means exhaustive) ways to theorise children and/as energetic phenomena: intractability/juxtaposition; embodiment; and speculative fabulation.

Intractability: juxtaposition-as-praxis

I began this book with two examples of intractable challenges facing children in São Paulo and Denmark. I did so for two inter-related reasons: to highlight that many of the challenges facing children around the world are so complex and multifaceted that, despite making enormous strides, interdisciplinary childhood studies *still* do not have the conceptual and methodological tools necessary to understand them and that, recursively, the conceptual and methodological arguments I develop in this book – and particularly modes of thinking and doing, *after childhood* – are not only conceptual or abstract concerns but are very much concerned with the challenges facing children around the world. That is not to say that this book has answers or purports to solve these problems: it does not. Yet it is concerned with developing some tools and, as I will re-state below, some *stories* that might at least help us to broach the complexity and intractability of those challenges.

That, indeed, has been one of the aims of this chapter, and especially of listing the variegated energetic properties emerging from our energy research in the previous section. For, rather than list off the inter-related challenges affecting youth in Denmark (connecting mental health to air pollution to transport), I have sought to juxtapose the forms of energy that flow through and characterise children and young people's lives across spatial and temporal frames. On an introspective level, doing so has been a rather fun game, if rather pointless (if, in other words, the only point was to create a list and to marvel at the sheer variety of different 'energies' therein). Yet whilst pointlessness, or *silliness*, may be an end-in-itself, it may be more-than-useful (Horton and Kraftl, 2005). The juxtaposing game is more than a game. For this practice could be both political and pragmatic: juxtaposing could be a vital form of practice, with *writing* (but equally, sketching or other forms of representation) a crucial mode of working, especially in interdisciplinary scholarship. It embodies but also extends beyond a particular form of *litanic* expression for thinking and doing, after childhood, which finds a rather different form elsewhere in this book, and especially in the next chapter.

For instance, Vitória's discussion of different cooking fuels evokes contemporary debates about cooking over open stoves in indoor domestic settings, especially in sub-Saharan African settings. Notably, whilst there has been little research on children and energy in the Majority Global South more broadly (e.g. De Hoop, 2017; Lusinga and de Groot, 2019), there *has* also been increasing attention to a range of more context-specific concerns. These include the entwined social and economic roles of children in *collecting* or generating energy (e.g. collecting firewood [Tian, 2017; Levison et al., 2018) and, in contrast with Vitória's positive view of the tastiness of smoked food, the (often negative) health and social effects upon children of repeated use of fuel-based energy sources in the home, which, in turn, are often compounded by gender and relative poverty (Biran et al., 2004; Mills, 2016; Patel et al., 2019). At the same time, the use of multi-fuel cooking stoves – often relying on wood, coal or gas – is a result of a kind of energy latency or 'lock-in'. This 'lock-in' exists for both infrastructural reasons (the reliance of countries like South Africa, Malawi and Lesotho on fossil fuels for energy) and cultural (values and tastes associated with certain kinds of cooking techniques). In turn, this sense of 'lock-in' resonates provocatively with notions of indenture in the Blood Bricks project discussed earlier.

Clearly, the example of cooking stoves illustrates the need not only for multi-dimensional but multidisciplinary research about children and energy. This is all the more pressing given that the references I cite above tend *either* to deal with questions of energy use, or health outcomes, or values, but, seldom, all three. I turn in Chapter 8 to the kind of multidisciplinary research that might be required to witness these forms of complexity including, but also *beyond*, the praxis of listing. But, here, I simply note that such forms of listing may, if nothing else, be the starting point. They are, at the very least, an *acknowledgement* that articulations of children with technology are complex and exceed our ordinary ways of thinking and writing about them, which tend to be confined to (environmental) education, technological fixes and (domestic) energy consumption. Praxes of juxtaposition – of listing, of writing, of sketching diverse energetic properties – can offer *openings* onto energetic worlds that are often hidden, and onto energetic processes that can (seriously) affect children's lives and their well-being, but in which children may not always be placed front and centre.

Embodiment: metabolism and expenditure

If juxtaposition is a praxis that might offer a starting point for theorisations of childhood and energy, we should not fall into the trap of simply allowing the effects and affects of those forms of listing wash over us. Rather, looking at the list above – and upon the more detailed examples on which the list is based – it becomes evident that certain forms of energy come to dominate – perhaps surprisingly, perhaps not so. In this chapter, and certainly counter to most energy literatures, it is *embodied* energies that dominate, even as these are entrained in complex, even intractable relationships with the timings of factory life or the exigencies of life at a school for excluded teenage boys. How, then, at the same time as retaining a commitment to juxtaposition and complexity might it be possible to nonetheless isolate and theorise diverse forms of embodied energies? Or, as Philo et al. (2015) put it in their explication of the embodied and affective doings of yoga: to which 'other energies' might we become attuned?

One answer is that such forms of 'isolation' can and should only be partial, and that theorising embodied energies must, equally, be a process of juxtaposition and compilation – again, of various approaches to children's embodied energy that, perhaps ironically, focus on both children or energy, one, or neither of these terms. Here, it might be interesting to juxtapose and theorise from a range of literatures. It could be instructive to begin with recent critical, feminist and materialist scholarship on 'fat', which has, for instance, revealed emotional and embodied dynamics of children's bodies towards the mobilisation of alternative physiological knowledges (Land, 2017). In her study of school-based physical education, Windram-Geddes (2013), for instance, explores girls' encounters with both material objects such as swimming pool water and 'non-material' discourses, conversations and idea(l)s surrounding body shape, in turn situated within wider societal talk about 'health' that articulates a 'fear of fatness'. Such intra-active encounters reveal new ways of understanding how girls and young women come to feel their bodies inside of contemporary obesity debates. This concern with embodied energies of food was expressed by Brazilian young people, but also in a rather different way by teachers at the school in Balsall Heath, who were worried at the combined effects upon boys of lack of sleep, constant use of computer games and over-reliance upon energy drinks.

Thus, it might, with care, be valuable to theorise the differently scaled *metabolisms* and *expenditures* that offer a rather different view of resource-power – of the imbrications of food and energy that I discussed in Chapter 3. By metabolism and expenditure, I am not solely interested in whether individual children are eating too much or not enough – although both of these have been shown to be the case in health research, whose underlying assumptions have in turn been critiqued by critical obesity studies scholars (Evans, 2010). Indeed, inspired by the kinds of juxtaposition cited above, I am not solely interested in the metabolism and expenditure of *food* but of food as just one set of energetic phenomena that flow through children's lives, environments and bodies, with or without others (whether electricity, fuel, sleep or whatever). We might then, if staying with food, consider approaches to understanding the increased 'energy-density' of foods consumed by children in especially richer countries and to efforts to reduce the energy-density of children's diets through, for instance, 'hiding' vegetables in meals (Spill et al., 2011; Robson et al., 2016). Significantly, whilst centred upon children, the notion of 'energy-density' (and, by contrast, energy-levity or energy-lack, both of which were issues for a subset of poorer young in our Brazil research, and for some of the sleep-deprived boys in Balsall Heath) extends beyond individual food practices to become a question of socialisation and affect. It is a question of socialisation at the scale of the family and particularly the intergenerational transfer of dietary norms (Robson et al., 2016). But it is also a question of socialisation in terms of the combined metabolism – the energy-density – of societies as a whole. Thus, not unproblematically, it becomes possible to 'measure' the metabolisms of communities or whole cities, as per recent models of 'urban metabolisms' that follow the energy and not the people – that follow flows of energy into, through and out of cities (Broto et al., 2012; McKinnon et al., 2019). Meanwhile, and perhaps more critically, and looping back to Windram-Geddes' (2013) work, it becomes possible to consider how energy density is narrated at larger spatial and temporal scales through technologies of affect – for instance, with respect to the obesity 'time-bomb' that may explode at some slightly undetermined point in the future if 'we' in the United Kingdom do not deal with obesity (and obese children) *now* (Evans, 2010).

Moving away from food, it should, too, become possible to witness ways in which energy is *expended*. Looking across disciplines – both within and beyond the social sciences – childhood scholars actually have plenty to say about energy expenditure. Significantly, this work is not only concerned with energy consumption (in the sense of buying or using energy or technologies that use energy) but the energies expended by children's bodies. In a striking piece inspired by French spatial theorist Henri Lefebvre, Kullman and Palludan (2011) undertook rhythm-analytical, ethnographic sketches of children's urban mobilities. They attended to multiple energies, of which children's energy levels were only a part: to "shifting environmental, social and technological rhythms – including urban traffic, weather changes and the affective fluctuations of tired and energetic bodies" (Kullman and Palludan, 2011: 347). These juxtapositions of matter and (energy) flow are redolent of the multiple energy registers we encountered in the nursery and, especially, the oddly rhythmical nature of our experience at St Paul's School (in, for instance, the contrast between the walk and the Jenga house). They also evoke the entanglements of food–water–energy in the lives of young Brazilians. We might, even, theorise all of these juxtapositions and registers as *energy assemblages*: in the case of Kullman and Palludan's work, these assemblages are patterned by shifting rhythms through which children gain mobilities-skills.

Although mindful of the critiques of health and medical scholarship about children's bodies – especially when it comes to obesity – physiological research about the expenditure of energy by children could also form part of the picture when it comes to theorising embodied energies. Indeed, drawing these approaches together towards more 'radical', interdisciplinary childhood scholarship might be a critical step in theorising such embodied energies. Specifically, it might be worthwhile to incorporate insights from a range of studies about the energies expended by children as they *move* with the (arguably more metaphorically articulated, although no less 'real') energies accounted for by Kullman and Palludan (2011) and the many other scholars of children's mobilities. For instance, in one of many similar studies, Bourdier et al. (2018) investigated differences in metabolic and fatigue profiles between 'untrained' boys aged 10 and 11 and both 'untrained' and 'well-trained' male endurance athletes. Strikingly – although in an admittedly small sample – they found that on balance, "prepubertal children were observed to be metabolically comparable to well-trained adult endurance athletes, and were thus less fatigable during high-intensity exercise than untrained adults".

A study by Kung et al. (2015) is emblematic of a rather different approach to energy, which focuses on bodily kinetics. Focusing on differences in joint kinematics between children who walk barefoot and wearing shoes, they measured a number of movements, including "[m]aximal joint displacements, maximal joint moments, angular impulse and mechanical energy at the hip, knee and ankle were analysed" (Kung et al., 2015: 95). There were key differences in joint flexion, extension, rotation and *energy generation* between the two styles of walking. For instance, walking with shoes prompted greater flexion (joint bending) of the hips, knees and ankles, alongside energy generation and absorption at the hips and knees. Meanwhile, walking barefoot generated a range of different forces, with "greater maximal hip extension and internal rotation . . . and energy generation by the ankle plantar flexors and invertors" (Kung et al., 2015: 95). The findings have implications for both shoe design and – perhaps somewhat surprisingly because walking may appear 'natural' – for learning an appropriate walking gait. For instance, Kung et al. (2015) suggest that a reduction in stride length may attenuate the braking forces associated with barefoot walking, and thus the stresses placed on particular joints in terms of energy absorption. However, they do not suggest that wearing shoes is necessarily 'better'; rather, minimalistic shoe designs may be preferable in order to ensure the development of foot structures as a child grows.

Looking beyond the specifics of these studies, and studies like them, and reading them in juxtaposition with the more 'critical', social-scientific accounts of children's embodied energies highlighted above, there are three implications for theorising complex and diverse energetic phenomena, after childhood. First, that there exists plenty of 'health' scholarship that is not concerned with diet, obesity or simply *increasing* children's activity levels. Although it might seem as if these are the principal preoccupations of child health researchers and policy-makers (and although there might in some cases be good reasons for these), it is important to note that health scholars also address a range of *other* concerns when it comes to children and energy. The first implication is, then, that these kinds of studies could help extend the purview of social-scientific studies of energy by posing questions beyond diet, obesity and activity levels; indeed, it is notable that, cutting across the hyperbole on the latter, untrained children performed virtually as well as trained adult endurance athletes in a number of tests in Bourdier et al.'s (2018) study.

The second implication is that these studies do not just provide points of empirical (or moral) inspiration but methodological and conceptual tools that – when used *carefully* and *critically* – may be integrated with more conventional social-scientific techniques in order to further the development of post-qualitative (Wolgemuth et al., 2018; Nordstrom, 2018), post-phenomenological (von Uexküll, 2013; Bogost, 2012; Ash and Simpson, 2016) and/or biosocial (Youdell, 2017; Osborne and Jones, 2017) methods. This is something that I pick up in far more detail in Chapter 8. However, in anticipating that discussion, I note for now that the studies by Bourdier et al. (2018) and Kung et al. (2015) offer a range of techniques and technologies that could extend the ways in which 'we' social scientists think children and energy: from ways to account for energy expenditure *beyond* what are actually quite particular forms of 'consumption', or 'urban metabolisms', to alternative schema for 'diagramming' bodily performances, kinetics and kinematics that extend beyond ethnography, sketching and mapping (Kullman and Palludan, 2011; also McCormack, 2003).

Thirdly, however, the studies of kinematics and walking gait also raise a series of unanswered questions that 'we' – as social scientists of childhood – might pose more routinely. These extend beyond the (equally important) considerations of the implications of these two studies for children's health, to the entanglement of health concerns within what I have termed the 'intractable challenges' that children face. Specifically – and here it might, after all, be helpful to return to questions of social difference and children's agency – *who are* the children who took part in these studies? By this I mean not only the sample ages or genders of the participants but more about their *geographical* context and the socio-material contexts of, for instance, barefoot walking. It is notably, for instance, that aside from research in India, the predominant contexts for similar studies to Kung et al.'s are all in the Minority Global North (Franklin et al., 2015). We can only surmise, but in many of these contexts barefoot walking is likely to be a choice; however, in the contexts of those children involved in collecting firewood or cooking in the home (cited above), walking barefoot may not be a choice and may be entangled with other concerns in terms of the intractable challenges they face, including the challenge with which I began this chapter: climate change. The question, then – which is conceptual, methodological and applied – is how future childhood scholarship might witness complex and multifaceted energetic phenomena that cut across questions of both health *and* social equality?

Storying: speculative fabulation

At the end of the previous chapter, I asked whether the accounts of archaeologists might – in a generative and certainly not dismissive sense – be considered, in part, as forms of speculative fabulation (following Haraway, 2011) that are ethically and politically attuned to larger temporal and spatial scales. At the end of this chapter, and cognisant of the infra-generations involved in Birmingham's energy histories in particular, I want to re-visit this term in order to consider a final resource for conceptualising energetic phenomena, after childhood. The inspiration for doing so sits firmly within our work in Birmingham, and the various acts of imagination, storying, play and, even, 'silliness' that were evident: connecting rock samples with the game Fortnite at the Lapworth Museum; the apparently destructive act of the Jenga house tornado; talking colonialism, and Trump, and climate change–related migration whilst doing a giant jigsaw map of the world; the energy superheroes exercise.

I am not naïve enough to think that these examples are necessarily either replete with the kinds of decolonising impulses or care for future (and past) socio-ecological generations that Haraway (2011) or other critical race or queer theorists embrace. There are traces, implied, perhaps unintentionally – although to claim this might in turn be to downplay the intentions of the young people involved. Some of these acts were, nonetheless, "seriously *playful* and so curious, inquisitive and risky" (Haraway, 2011: 6). The Jenga-tower tornado was, for a moment, all-consuming; the energy superheroes generated 'affective energies' (Philo et al., 2015) in which the children and we "palpably [ran] the risk of coming to care about, even to love, the fabulated" characters of their making (Haraway, 2011: 6). By dressing-up, making unexpected associations, telling stories, being 'silly' and 'disruptive', children and young people in Balsall Heath articulated manifold, ephemeral, technocultural assemblages. Rather than provide solid answers to the Anthropocene predicament – although acutely aware of our many shared problems – they responded with provisional acts of fiction; with multiple lines of flight (Deleuze and Guattari, 1988), many of which headed simply nowhere. Yet to apparently head 'nowhere' is, precisely in the grand utopian tradition, an opportunity for disruption (Kraftl, 2007); an act that may unsettle the single-minded technofantasies of hard engineering solutions to climate change or habitat destruction. This is a gentle, playful act of claiming that does not "*point* toward future utopia or dystopia" (Haraway, 2011: 6; my emphasis) but towards a disposition to care, somehow, in a way that is "wet, emotional, messy", where

> the past, present and future are all very much knotted into each other, full of what we need for the *work* and *play* of naturalcultural restoration: less deadly curiosity, materially entangled ethics and politics, and technical and organic well being.
>
> (Haraway, 2011: 6 & 17)

These acts of care – these traces – felt speculatively fictional because of the ways in which children and young people wove 'technologies' and media ecologies (Bryant, 2014) into the stories they told, the scenarios they imagined, and into their embodied, messy, playful performances.

Interestingly, many of the modes of expression of some of the teenage boys at St Paul's School would afford a sense of *not* caring. One might also question the extent to which the energy superheroes work, or the playful moments enjoyed whilst splattering paint, might genuinely foster the kinds of decolonising, caring ethics that enable us "to move toward multi-species reconciliation" (Haraway, 2011: 8). Yet, as Taylor (2019: 5) and others remind us, the effects and affects of "[h]eading off leftfield with minor players" (also Katz, 1996) are not simply escapism or a reconfiguration of the power relations within the production of knowledge by including those actors deemed 'irrelevant' (and, being provocative, which group is seen by society as *less* relevant to environmental discourses than teenage, ethnic-minority, boys with social, emotional and/or behavioural differences?). Rather, they are acts of "scaling down, slowing down and attending to the seemingly 'obscure matters' of these assumed-to-be insignificant minor players [. . . whose] relations with wildlife [can be] so illuminating" (Taylor, 2019: 6). These stories – these acts of playful fiction – are forms of 'passionate immersion' (Tsing, 2010: 19) and making kin (Haraway, 2016) – not only with non-human animals or plants but with technologies, rocks, paints, clothes and far more besides. These are speculations with diverse energetic phenomena.

Finally, perhaps it does not matter (for now) whether the energetic phenomena emergent from the speculative fabulations in Balsall Heath mirrored precisely the forms of decolonising, caring ethics that Haraway (2011) calls for. They took place in a context and format far away from the artistic practices in Australia she discusses, and – without at all wishing to speak for Haraway's intentions – I suspect that she would be the last to efface the specificities of place in efforts to multiply, rather than to reduce, 'other energies' in responding to the Anthropocene (or

Cthulucene). What is worth remembering, however, is that through the odd, play-ful, perhaps 'silly' acts of fiction we encountered in Birmingham, juxtapositions of diverse energetic phenomena – of real concern to *these* and *other* children, if the #climatestrikes were anything to go by – were nevertheless narrated. This is why, for instance, newly planted urban trees are 'gas-guzzling mother-fuckers'; why bricks simply *mattered* to teenage boys growing up in *Birmingham*; and why the children at the playscheme chose to leave their batmobile behind when they went off to fight crime. What, after all, could be more eco-friendly than that?

Conclusion: energetic phenomena, after childhood

The inclusion of popular cultural references, such as batmobiles, *within* the kinds of 'minor stories' that Taylor (2019) and others advocate is a reminder, if one were needed, that we need to 'stay with the trouble' in all sorts of ways when it comes to working through, rather than simply celebrating, the responses of any and all humans to current environmental challenges (Haraway, 2016). It is tempting – particularly given how I draw in part upon the work of Bryant and Bogost in this text – to be taken in with the media- and machinic-assemblages to which some of the 'small stories' in this chapter are coupled, and which, as I argued in Chapter 5, circulate massively in the contemporary world. Nevertheless, these stories raise questions about the extent to which such popular-cultural imaginaries, and the global consumer systems in which they are embedded, might offer generative, progressive, alternative or – depending on one's point of view – 'legitimate' fod-der for challenging technocentric responses to the Anthropocene. Importantly, as I was at pains to point out in Chapter 2, these extend beyond and highlight the limitations of new materialist and object-oriented thinking; or, at least, the need to situate them more explicitly within other concerns and frames (whether popular cultures in this chapter, or the 'cuts' in the previous one). Certainly, it is important to recognise that many of the alternative educators and activists with whom I have worked in the past (see Kraftl, 2013b, 2015) would frown upon at least some of these stories; yet others would see opportunities for working with and against the grain of games like Fortnite, opening up what Holloway (2010) calls 'cracks' (perhaps 'cuts'?) in the capitalist system. This impulse does, perhaps, run a little counter to the minor theories and stories of feminist-inspired work that seek to work outside, rather than with/against 'the master's tools' (Taylor, 2019); yet it may also enable an acknowledgement that there are multiple, legitimate ways of (not) engaging with global-capitalist systems that would benefit from a critical-comparative perspective (see, especially, Gibson-Graham, 2006).

These specific reflections upon the appearance of popular-cultural (and very much of-their-time, as I wrote this book) 'machines' in some of the stories in this chapter are included here for two reasons: first, in order to briefly pick up the thread of the argument developed far more fully in Chapter 5, about the role of digital media and machines in thinking and doing after childhood; second, as an entry-point to a wider set of reflections on the ways in which I have tried to theorise diverse energetic phenomena and childhoods in this chapter. I began the chapter by briefly revisiting notions of resource-power that were introduced in Chapter 3, because a key argument in this chapter has been that theorisations of energetic phenomena should not merely be taken up by a fascination with the material (or, rather, immaterial) properties of energy that make it qualitatively different from some of the other material phenomena discussed in this book and, indeed, by many contemporary childhood scholars. Rather, I have argued that any consideration of energy should be attuned to the very same questions of *power*, *equality* and *justice* that should undergird any consideration of resource nexuses and digital media, as well as remaining attendant to material things (like bricks). Specifically, I sought to call out the significance of intersecting identity traits as these play with/in energetic phenomena: of what might be termed extra-sectional social-material processes (Horton and Kraftl, 2018). These processes came through most evidently in my discussion of our work at St Paul's School, but also in the rather different experiences of Adriana and Vitória, given the social and geographical contexts in which they live in Brazil.

A key manoeuvre in this chapter has been to tell a range of smaller stories and engage with speculative fabulations (Haraway, 2011) that could enable more diverse, complex accounts of children's entanglements with and as energetic phenomena. The chapter accounts for a range of what, in their call for 'new energy geographies', Philo et al. (2015) term 'other energies' – both in terms of the kinds of concerns that are the usual substance for energy (as resource) researchers, and the preoccupation with technology, education and (domestic) consumption in the fairly narrow literature on children and energy. Indeed, our work with children and young people, across a range of sites and contexts, gave rise to a bewildering array of energetic phenomena that surpassed even the categories conceived by Philo et al.: material, embodied, emotional, mediated, technological, mobile and more besides. In turn, apparently small stories articulated with 'larger' ones: from Birmingham's energy histories (and the expression of those latent histories in its brickwork, dismantled railways and canal systems), to extreme weather events, to the organisation of time in the modern factory.

Finally, I have attempted – provisionally – to formalise these 'other energies' into a theorisation of children's engagements with and as *diverse energetic phenomena*. Compared with other chapters, in particular, this has meant foregrounding children's own actions (and voices) to a greater extent, whilst recognising that, as I have argued, energetic phenomena are (literally) harder to grasp, elusive and shifting. This theorisation has involved three elements. First, and underpinned by a commitment to scrutinising power and (in)justice, by a sense in which any consideration of energy/ies must see it/them as entangled in a series of *intractable* challenges that matter, profoundly, to children's lives. These challenges may be narrated as a 'nexus', or, simply, as a series of compound processes that may pattern and be patterned by the relative social and geographic positioning of the children who are implicated in them. As I have argued, *one* response – which may fall woefully short of the mark in actually addressing such intractable challenges – is to engage in a *juxtaposition* of different energetic phenomena, perhaps via *listing* (Latour, 2005). This book is full of other possible and perhaps equally provisional responses to the question of intractability, in which children and childhoods are often decentred or more out of focus. Second, I have promoted a particular attentiveness to 'other energies' that remain hidden in energy scholarship and policy-making, whether with children or not – and especially *embodied* energies. It may be that, as Philo et al. (2015) observe, such energies are not always felt or named as such – or that scholars of energy resources would place them in the same register as the kinds of energy derived from fossil fuels or renewable energy sources. However, the point is that whether metaphorical, felt or material, these 'other energies' are not distinct from but are entangled with, productive of and constituted by those more 'familiar' energy (re)sources. Therefore, as I argue in the next chapter, more 'radically' interdisciplinary modes of thinking and doing childhood research are required that might critically analyse those multiple and complex forms of energy. Specifically, I asked how childhood scholars might use techniques from research about human energy expenditure and kinematics to move beyond a preoccupation with obesity and physical activity, and how in turn an awareness of the social positioning of children who walk barefoot might intersect with an analysis of the other related challenges those children face in collecting firewood or mitigating the effects of climate change at home.

Third, I ended the chapter by considering whether and how the stories told and performed by children in Birmingham and Brazil might usefully be conceived as forms of *speculative fabulation* (Haraway, 2011). I argued that, whilst living in a very different social and geographical context from Haraway's work in Australia, children and young people were involved in acts of reflection and imaginative, playful world-building that were sometimes challenging, sometimes silly, but also, sometimes, at least implicitly *caring* about the energy histories of previous generations in Birmingham, and what might need to change in future generations. Moreover, speculative fabulation provided them – and me – with a process of narration through which diverse energetic phenomena could be juxtaposed. Consequently, I argue, the three notions of intractability, embodiment and speculative fabulation could form additional strands for thinking and doing, after childhood.

Note

1 *Peaky Blinders* was a popular UK television drama series set in 1920s Birmingham, which charted the fortunes of a violent gang as they sought to gain power and wealth in the industrial city.

8 Synthesis and stickiness
Lives of plastics, metals and other elements

The Great Burger King Meltdown; or, starting with plastics

On 19 September 2019, the fast-food chain Burger King announced in a major press release that it would recycle children's unused toys for free in the United Kingdom. In a series of tweets to its 48,000 UK followers, it announced that, in order to address issues of plastic waste, it would not only stop giving out free plastic toys with its meals, nor only recycle their own toys, but would take any plastic toys brought to their stores. Most toys cannot be recycled with household waste and therefore often end up in landfill – or else appear discarded in apartments or oceanic trash vortices or elsewhere (Chapter 4). Therefore, Burger King took the decision to work with a company that literally melts down the toys and turns them into other products that the restaurant chain will use. According to the manager of the plastics company, the process will not release any harmful gases into the atmosphere and will reduce energy consumption by 88% over the manufacture of 'new' plastics (BBC News, 2019b).

Aside from the BBC's reporting, Burger King's own coverage of the Meltdown on social media makes little mention of what – or rather *who* – instigated this move. It transpires, though, that the chain's decision was influenced by an online petition by two British girls. Sisters Ella (9) and Caitlin (7) McEwan secured more than half a million signatures for their plea to several large fast-food companies to stop giving out free toys and to consider the effects of plastics on the environment.

However effective Burger King's decision might be in reducing plastic waste, the Meltdown offers a window onto what – riffing on the focus on plastics – might be termed the *synthetic* nature of childhood. By synthetic I do not mean (or only mean) the more familiar, pejorative discourse about childhoods as somehow artificial, toxic (Palmer, 2015) or disappearing (Postman, 1985). Rather, just like plastics, childhoods and – because of their perceived or actual malleability – children can easily be attached to other stuff. There exist infinite chemical combinations for the synthetic polymers that make up plastics, and, perhaps more importantly, despite the commonsensical notion that nothing *sticks* to plastics, they nevertheless become entangled with, lodged in and attractors for a vast range of non-plastics, at scales ranging from the nano to the macro. Elsewhere in this book I have argued for a sense of the 'pull focus' – wherein childhoods, children and an attentiveness to them are simultaneously decentred and recentred. In a sense, the same is true in the above example: the focus is on plastics, foremost, but those plastics happen to *matter* because they are *toys*, and because of their inevitable association *with* childhood (although at most major chains, adults can purchase 'kids' meals' and receive a free toy, too). And, I would contend, the narrative of the BBC's article is all the more powerful for the 'big reveal' – how the article starts with Burger King's decision but only – and apparently quite knowingly – highlights that the move was the result of a petition by two girls halfway through. This, in microcosm, is exemplary of the political and affective power of the child-hood 'pull focus', and not least because of the ways in which plastics and child-hoods are *synthesised* with and *stuck* to one another.

The figurative power of the 'Meltdown', though, offers up a different view. As a play on words, it offers humorous connotations of that classical expres-sion of childhood anger: the tantrum. Indeed (and this is observation is not aimed at Burger King themselves), one of the ways in which generally conserva-tive, climate-change-denying adults have sought to undermine the #climatestrike movement and its young leaders is to dismiss it as an insubstantial, fleeting, child-ish outburst. As many commentators have argued, however, this kind of anger is not insubstantial and, although it may not (at the time of writing this book) have yet offered concrete 'solutions', is a necessary part of an activism designed to shake the world out of its plastic-habituated apathy (on emotion and activism, see Brown and Pickerill, 2009). The 'Meltdown' is, then, to be understood as an appropriately measured response to the urgency and seriousness of the plastic moment in which we find ourselves. And, in this move, the coupling of plastics to *childhood* – that stickiest of concepts – is a powerful one. It has as much traction as, say, the classical couplings of children with fears about public space, or about (digital) technologies.

Yet these kinds of discursive attachment or *stickiness* only tell part of the story. The idea of the Meltdown reminds us, as I intimated above, of the malleable and therefore *synthetic* nature of plastics (and, as I will demonstrate later, of child-hoods). Crucially, and as I will be at pains to argue in this chapter, we might *start with* plastics – but they are not the be-all, the end-all or the after-all (see Chap-ter 1). Thus, this chapter is not only about plastics – indeed, it is not really about plastics at all. This is for two very good, inter-related reasons. On the one hand, plastics only represent part of the problem; for many environmental campaigners, the current fervour around plastics constitutes a distraction from other major issues such as climate change, water and air pollution (BBC, 2018a). On the other hand, plastics are, in turn, entangled with/in those issues: they are part of and may exac-erbate water pollution, for instance, even if the effects of at least some plastics on the health of humans and other species may not be as serious as those of other chemical pollutants. As a sticky problem (or, following Morton [2013], a viscous one), plastics offer a way in and, both literally and metaphorically, a 'hook' for (re)thinking childhoods as synthetic and sticky. But that is all; as I will show, there are many other matters, besides plastics, which are equally pressing, gelatinous, cloying and treacly. We need to *stick* with the trouble (compare Haraway, 2016).

The first step for doing so *is* to start with (the 'problems' of) plastics themselves – both in terms of their own synthetic make-up and their variable abilities to stick (or not) to non-plastics. This step offers a reminder of OOO and the notion of the 'interface' that I introduced back in Chapter 5 when discussing digital media. It also offers an opportunity to introduce Bogost's 'alien phenomenologies' and to link again with a range of critical race and queer scholarship around materiality and environmental (in)justice, which, in combination with theoretical frames introduced throughout the book, provide an analytical framework for thinking synthetically, and stickily, after childhood. The remainder of the chapter is organised around different 'cuts' through the Plastic Childhoods project, in which, in anticipating the book's final chapter, forms of *silliness* and *trauma* interface in all kinds of ways. Firstly, I start again with plastics – this time through the example of some totem poles that were created during the Plastic Childhoods project, and in which the inter-relating aesthetic and material properties of different plastics were tested out. Inspired by Bogost's work and returning to the notion of hyperobjects, the third section of the chapter moves deliberately away from plastics to account for the appearance and circulation of a range of metals and other elements in children's lives. The final section of the chapter takes seriously a plea to stick with children and childhood – and with more traditional approaches for listening to children's 'voices'. However, drawing on a project that combined social science, art and environmental nanoscience, the chapter also makes and exemplifies a case for perhaps more 'radical' forms of *interdisciplinary collaboration* – what Belontz et al. (2019) term a 'synthetic collective' – in thinking and doing, after childhood. In doing the above, and in (as far as is possible) drawing together some of the threads developed elsewhere in this book, this chapter also functions as a kind of ante-'conclusion', because the final chapter articulates a still wider set of questions and concerns.

Starting with plastics (I): the problem with plastics

The problem with plastics is, in part, a problem of (and for) the concept with which I ended the previous section: interdisciplinarity. Plastics represent what – in the fashionable parlance of 'global challenges' research landscapes – are termed 'wicked' (Belontz et al., 2019) or 'nexus' (Chapter 3) or 'intractable' challenges (Chapter 7). Simply put, this is because of the sheer scales (both infinitesimally small and unimaginably huge) and complexity of how plastics have become 'the substrate of advanced capitalism' (Davis, 2015: 348). These scales and complexities relate directly to plastic's synthetic and sticky qualities, but also a lack of knowledge about plastics that scientists are currently seeking to address.

Notwithstanding that lack of knowledge, it is commonly accepted that the production and wastage of plastics have been increasing rapidly. Although Bakelite – the earliest form of plastic – was invented at the turn of the twentieth century, it was only from the 1940s that the production and consumption of plastics began to escalate. Analysis of plastic fragments deposited in the Santa Barbara basin shows that the build-up of plastics has roughly doubled every 15 years since the 1940s (Klein, 2019). There is also greater variety over time – in terms of both the diversity of plastic polymers and objects themselves, ranging from clothing fibres to resealable ('Zip') food bags to microbeads (on which more later on) (Klein, 2019; Novotna et al., 2019).

The scale and complexity of plastic proliferation is often taken as a key indicator for evidence of the Anthropocene, given the presence of plastics as a 'horizon' in water, air and rock (Zalasiewicz et al., 2016). As plastics originate in, impinge upon but also exceed human agency and perception, the allegory of a 'horizon' is an appealing one: it speaks of the phasing and temporal undulations of hyperobjects like plastics (Morton, 2013), as "vectors along which a world is unfolding" (Bryant, 2014: 124). Let us take each of these vectors (water, air, rock) in turn. Perhaps the majority of scientific research has focused on water. In terms of scale, the best estimates are that between 4.8 million and 12.7 million tonnes of plastics are dumped into the world's oceans each year (Jambeck et al., 2015). In addition, between 0.49 and 4.6 million tonnes reach the oceans via the world's rivers as both macro- and microplastics (Schmidt et al., 2017). As a result, microplastics (<5mm in size) and nanoplastics (<100nm) are recognised as a globally circulating pollutant; although the health effects of plastics borne in water (as opposed to other forms of exposure) are still unclear, a few studies have focused on a comparison between treated and untreated water (Novotna et al., 2019). It appears that untreated water has much higher levels of microplastics (up >4,000 pieces per litre) whereas lower levels have been found in drinking water, with variations between samples from different countries probably explained by different water treatment practices (Novotna et al., 2019). As has been widely reported in the media, bottled water (including that in glass bottles) has also been found to contain plastics – likely down to bottling techniques – often at higher levels than treated tapwater (Koelmans et al., 2019). The most common types of plastics found in drinking water are polyethylene (PE), polypropylene (PP), polystyrene (PS), polyvinyl chloride (PVC) and polyethylene terephthalate (PET); the explanation for their prevalence in water sample is partly down to their popularity in the manufacture of consumer products and partly their density (Koelmans et al., 2019). Thus, connecting with analyses made in detail elsewhere in this book and this chapter, the presence of plastics in drinking water *is* down to human action – we manufactured and released these plastics into the environment after all. However, it is *also* a result of the properties of the plastics themselves once they are released and of all intents and purposes that exceed human control – and, more to the point, of the ways in which they *interface* with/in the properties of water given their relative size, density and shape.

A similar point about the interface or encounter can be made in terms of recent evidence about the presence of plastics in air and rocks. Herein, even if for Harman (2011b: 150) much of an object recedes, objects also allow us (humans and other objects) to "bathe in them at every moment". Some properties remain perceptible or workable to the other objects with which they come into contact, and that perceptibility or workability (and the opportunities for synthesis or attachment) will depend on how each object encounters the other. Thus, "[w]hen they [objects] interact through vicarious causation, they do so . . . in relation to the qualities in which they 'bathe'" (Bogost, 2012: 66). Exemplary of these forms of vicarious causation from which humans recede or are absent are the ways that *particular types and forms* of plastics are transported in the atmosphere. Dris et al. (2016) demonstrate that the properties of microfibres (for instance, from synthetic clothing) afford their movement over long distances – hence why plastics have been found in remote areas of the world far from human settlements. A fascinating study (Bergmann et al., 2019) found that levels of microplastics in Arctic snow near Svalbard were, although lower than those found in the European Alps, nevertheless significant. The presence of microplastics (and other matter, to which I will return below) was thanks to interfaces occurring at the micro- and intercontinental scales: microscopically, as it forms in the atmosphere, snow has an ability to bind particles of matter other than water into snowflakes; intercontinentally, Svalbard sits at the confluence of several long-distance airflows, driven by different pressure systems from North America, Europe and Siberia.

The encounter between plastics and rock is very different. A case that has attracted attention amongst scientists and artists alike has been the discovery of so-called plastiglomerate on Kamilo Beach, Hawaii, in 2014 (Huang, 2017). This rock, formed of lava, sand, beach materials and melted plastic (an earlier 'Great Meltdown', perhaps?), has been treated as a rock sample, a new kind of stone and a kind of readymade sculpture (Corcoran et al., 2014; De Loughry, 2019). It has also been considered an exemplar of the Anthropocene condition given the entanglement of plastics with and as *rocks* – substrate – and not 'merely' in oceanic trash vortices. Importantly, the term 'entanglement' – a favourite of new materialist thinkers, especially writing on childhood, and a word that I have deployed liberally in this book – does not quite cut it in this instance. Rather, the fact that the plastic has *melted* is a literal embodiment of, and metaphor for, a slightly different encounter: one of *synthesis* (where rock, sand, beach matter and plastics become-one-another in an ostensibly new, singular rock form) and *stickiness* (where heat has caused lava and plastic to melt into one another, and cooling has meant that they remained fixed-as-one through a certain viscosity). I will come back to the (vital) point about matters other than plastic in a moment.

Currently, a fair amount of scientific and critical conceptual attention has been centred on the risks to (especially human) health of exposure to plastics of different kinds. *Some* plastics – and the plasticisers used to produce them – have been associated with a range of ill effects, particularly in children and young people, including obesity, miscarriages, feminisation of male foetuses, early onset puberty, reduced brain development and a range of endocrine disorders, because some plastics can work as an endocrine disruptor (Davis, 2015; De Loughry, 2019; Belontz et al., 2019). These effects are often geographically and socially specific and associated with spending significant amounts of time handling and processing plastics without protection, especially in conditions (such as recycling plants) where plasticisers may be released causing high levels of exposure. The most famous of these places is Wen'an, a village devoted to recycling (often Western) plastics, which has been called a 'dead zone' where people and landscapes have been subjected to the 'slow violence' of decades of exposure to plastic(iser)s (Davis, 2015; Davies, 2019). This leads to an open question as to whether people and places like the waste recycling facility at the eco-park in Guara (Chapter 3) might be equally exposed and, more generally, the many children who work, often unprotected, at waste disposal sites around the world (children that Cindi Katz [2018 provocatively terms the 'waste of the world').

However, other plastics may be perfectly safe – whether because they are effectively chemically inert or because, if ingested, they simply pass through the human body without causing any effects (breathing in plastic fibres may be marginally more dangerous, although these are usually expelled through coughing or sneezing) (WHO, 2019). Thus, whilst microplastics have been found not only in water but beer, fish and a range of other foodstuffs, and whilst small trials have found plastics in human stool (an average of 20 particles per 10 g of stool), the current guidance from the WHO (2019) and other scientists is that spending resources on removing plastics from food or water is not a priority (Koelmans et al., 2019). This is because, as the WHO (2019) puts it baldly, the risks posed by a range of bacteria and viruses – and by climate change and disasters – are better understood and therefore a priority, at least when it comes to safe food and water supplies. Quite where this leaves those (children) more directly exposed to plastics at Wen'an and other locations is an open question.

Despite the contemporary fascination with plastics, and despite the very real problem they pose, it is hard to speak of plastics alone. This is why I argue in this chapter that it is only ever acceptable to *start-with* plastics. There are various reasons for this – some of which I have already articulated – but it makes sense to bring them together to make the point more forcibly. Many of these have to do with plastic's involvement in processes of *synthesis* and *stickiness*. First, plastics are far from the only problem facing the planet and of human making. Plastics intermingle with and exist alongside climate change, habitat loss and myriad other forms of pollution. Indeed, as I will go on to show, not only are other elements – whose adverse health effects are far better known – more readily identifiable in children's environments and bodies than plastics but plastics may not be all bad. Plastics have differential value and – for instance, when it comes to preserving food – may prevent food wastage and associated carbon emissions. Furthermore, when (re)turning to children themselves, and their own experiences, it also becomes abundantly clear that plastics are co-implicated with all kinds of other material stuff in ways that can be judged 'good' as well as 'bad'.

Second, as the example of the plastiglomerates demonstrates, it is when plastics interface with objects after the human that they become more interesting. Again, this has everything to do with what Harman terms 'vicarious causation', where particular kinds of plastics and other particular materials, each with properties that present themselves (or 'bathe') in particular ways to one another, as they encounter one another, become synthesised or 'stuck' together. For instance: "Low-density microplastics found in benthic sediments [. . .] may result from the presence of mineral fillers, development of biofilms, adsorption of clay minerals, and flocculation with organic matter" (Belontz et al., 2019: 856). The wording chosen for the title of the source work cited as evidence for this assertion is illuminating – particularly from a speculative-realist perspective: "*Interactions* between microplastics and phytoplankton aggregates: Impact on their respective *fates*" (Long et al., 2015; my emphasis), because it attests to the interface between objects and affords them – through the use of the term 'fate' – a kind of vital materiality or actancy (Bennett, 2010; Latour, 2005).

Third, it is plastic's (perhaps unexpected) capacity for synthesis and stickiness that presents an arguably greater risk – at least to human and non-human ecosystems at large, beyond the specifics of plastic recycling plants. Put simply, organic toxins and pollutants may be adsorbed onto plastic surfaces, rendering them hosts for potentially harmful bacteria and other organisms – especially when combined with many other kinds of matter in oceanic trash vortices. Thinking infra-generationally (Chapter 6), the flotsam and jetsam crossing and combining in the world's oceans are not new, and nor is their capacity to carry all manner of 'nasties' – yet plastics offer a new-generation host or vector – and one whose *sticky durability* looks set to exceed the timeframes of *human* generations or, indeed, the human species itself. Thus:

> [t]hrough the distant past to modern times, these materials have also attracted a diverse biota of sessile and motile marine organisms – *freedom travellers* (hitch-hikers and hangers-on if one likes). This process has been a mechanism in the slow trans-oceanic dispersal of marine and some terrestrial organisms [. . .] the hard surfaces of pelagic plastics provide an attractive and alternative substrate for a number of *opportunistic colonizers*. With the quantities of these synthetic and non-biodegradable materials in marine debris increasing manifold over the last five decades, dispersal will be accelerated and prospects for invasions by *alien* and possibly aggressive *invasive* species could be enhanced.
>
> (Gregory, 2009: 2018)

For many critical queer and race theorists, plastics present *humans* with a twofold temporality that resonates with my theorisation of infra-generations. Plastics are 'untimely' because their effects may be delayed as their (partial) decomposition is slowed by lack of exposure to heat and light (the slow-mo version of the Great Melting?), whether because they are buried in the earth or found in the darkest depths of the ocean (Ghosh, 2019: 280). Thus – and here childhood appears in passing – Ghosh (2019) talks of how the remnants of a plane shot down 60 years previously, 9,600 kilometres away, were found in a dead albatross and how, at the Midway Atoll, millions of the same bird have been found to have plastics in their stomachs, including wrappers from children's sweets. At the same time, the production, consumption and divestment of plastics have led humans down a path of destruction and ultimate (species) death from which there is no escape, no outside. As Davis (2015: 351) puts it, plastics don't discriminate between class or geography:

> [p]lastic, in this sense, represents the fundamental logic of finitude, carrying the horrifying implications of the inability to decompose, to enter back into

systems of decay and regrowth. In our quest to escape death, we have created systems of real finitude that mean the extinguishment of many forms of life.

One way to interpret this sense of 'finitude', and one in line with the notion of infra-generation, would be to read it alongside its articulation by Meillassoux (2010; see Chapter 1) as that which is anterior or ancestral to human life and perception. In plastics, we see the early traces of our own death as a species – exemplified at present by the death of others, whether human or non-human. Literally, we catch a glimpse of life *after* the human, in which the traces of childhood in wrappers and toys are fully implicated (see Chapter 9). Another way to interpret it, which is equally valid, is to ask again – as we should, constantly, of the notion of the Anthropocene – whether, if plastics don't discriminate between class, geography or other forms of *human* difference, the difference that we humans can make is to do precisely that.

Thus, for instance, Huang (2017) makes a powerful argument for how the great Pacific Garbage Patch (the largest of the oceanic trash vortices) is racialised. And it is here that, notably, materialisations of *childhood* (attached to plastics, as I argued in the opening to this chapter) re-enter the equation. Once again, this is a case of starting-with plastics to arrive at a point where we consider *plastics-with-*: logics of attachment or stickiness that take us beyond (or after) plastics. Resonating with my discussion of energy in the previous chapter, Huang makes the argument that the Pacific Garbage Patch is not, actually, an object: rather, it is a gyre, a swirling mass wherein matter/s is/are caught up, synthesised and, sometimes, released. In a familiar Modern move, the fantasy is that the Patch is the Other to 'clean' Western, human lifestyles. But, following Barad (2007), Huang argues that the Patch is not other; it does not exist on the other side of an agential 'cut' – and here the (I think unintended) resonance with the archaeological interpretation of the term 'cut', introduced earlier in this book, is striking. This is stuff that is meant to be forgotten, so amorphous that the details or identities of individual 'objects' do not matter. Yet this stuff *does* matter to the encoding of Asian (American) racial identities – or, rather, to the encoding of Asian identities by White Americans. For, the Patch functions as a gyre of social-psychological attitudes to the Chinese as well as a gyre of material stuff: where 85% of all toys played-with in America are manufactured in China; where those toys are associated with cheapness and 'Chinese crap' (and hence throwaway – just like Burger King's free toys); where notions of the influx of Chinese products and the potential 'harm' posed by certain toys is associated with Asian immigration to the United States (also Chen, 2011); and where, although the Patch is viewed as extra-national, it becomes associated with particular national identities – ironically, with the Chinese industry, where much of the stuff also comes from the United States, the United Kingdom and other Western nations (Huang, 2017).

All of this – all of plastic's synthetic and sticky qualities – begs the question of how to respond. In this chapter, my response will be guided by two frames of reference that – in the spirit of another of plastic's properties (its stubborn ability to break down and not to wholly decompose) – do not fully combine or enmesh:

Bogost's alien phenomenology and critical race and queer theorisations of (especially chemical) matter. I discussed the latter frames at length earlier in Chapter 2, so I will not go into great detail in repeating their arguments here. Rather, I focus on how Bogost's work might add to and sit in tension with these and other conceptual tools I have developed throughout the book for thinking and doing, after childhood.

Alien phenomenology and after: (instrumental) interfaces, speculation, litany, ontography and metaphor

Ian Bogost's (2012) alien phenomenology offers a particular conceptual and methodological framework for understanding what Bennett (2010: 6) calls "thing-power". But, in a move akin to Von Euxkell's ethology, Bogost offers a sense of how non-human things *encounter* or, even, *perceive* the world. This is a step beyond ANT, because it does not represent a 'flat' (rather, what Bogost calls a 'tiny') ontology of relations but a way of reaching out to what it is *like* to be a thing, during an interface with other things. This is a step beyond post-humanisms (for Bogost, 2012: 7), because such approaches "still preserve humanity as a primary actor". Rather, in Bryant's (2014: 62) words, it is to "seek to determine the flows to which a machine is open, as well as the way that machines operates on these flows as they pass through the machine". Although we cannot experience the world as (say) a bat, or a doll (Chapter 4), it is possible to make all sorts of inferences in any attempt at determination. Von Uexküll's (2013) ethological response to this is to deploy a range of cutting-edge scientific techniques that afford a sense of how a bat senses: through instruments to detect ultraviolet light and emerging knowledge about optics and physiology, we now know far more about how a bat detects insects that it perceives as food.

The fundamental point here is that science – and what I term the 'instrumental interface' (measurement techniques for deriving the experience of things) – could be a *part* of the picture of, rather than an enemy for, childhood studies (after childhood), as it has so often been positioned. Indeed, I am drawn – with care – to the idea of using cutting-edge instrumentation to follow the presence, movements and lives of things as a way of better understanding how those things cross-cut with human (children's) lives in ways that profoundly *matter* to those lives (with the use of the API on an online selling platform in Chapter 5 being a rather different example). Just like writing about archaeology or energy elsewhere in the book, a dose of *speculation* is also required, as I will demonstrate – and here I am a strong advocate not for science for the sake of science but science in conversation with art, social science and experimental modes of writing. Thus, later in the chapter, I speculate about the lives of aluminium, titanium and zirconium, using samples analysed through the techniques of environmental nanoscience.

But Bogost's call to understand – or at least empathise – with how things are open to or perceive the world goes further and is also cognisant of these dangers. He argues that a reliance on science could resurrect a kind of "scientific naturalism [. . . where] never mind the *sort* of stuff [. . .] there is always *some* stuff out of which all others can be explained" (Bogost, 2012: 13). This reductive manoeuvre is one of which so many social theorists have been (rightly) critical. As a theorist of computers, videogames (Bogost, 2015) and much else besides, Bogost sees part of the answer in a detailed exposition of how machines like computers encounter the world. One strategy – which overlaps with Latour's notion of listing, discussed in Chapter 4, and the notion of the *litany*, explored otherwise below in relation to critical race theory – is to offer a multidinous, multidimensional answer to a question like "What is *E.T.?*" (Bogost, 2012: 17). Referring to a 1982 Atari videogame, Bogost offers a striking page-long, listed answer that opens out the different dimensions in which that question can be answered: from the kilobytes of opcodes and operands that drive processor operations, to the game's assembly code, to the "RF modulations that result from user input and program flow", to the "moulded plastic cartridge held together by a screw" (the game cartridge inserted into the console) that "reveals the chip's contacts when actuated" by the console, to the experience of playing the game, and so on (Bogost, 2012: 17–18). I attempt something similar to this approach when writing about aluminium, later in the chapter.

Very much related to his attention to listing, Bogost (2012: 50) offers "*ontography*" as a way of "cataloguing things, but also drawing attention to the couplings of and chasms between them". One example Bogost (2012: 51; original emphasis) gives of ontography is the "*exploded view* diagram", which one might find in instructions for LEGO™ or in Leonardo da Vinci's notebooks or in car parts manuals, showing the ways in which screws, buts, bolts, washers, springs (etc.) fit together in any given assembly. Another example is a particular kind of photography, and particularly the work of American artist Stephen Shore, who documented ordinary aspects of American life in a way that attempted to allow units in any 'scene' – such as the burger, bun, plate, relish, table cloth and drink on the table in a diner – to "reveal themselves [. . . exemplifying] the ways that human intervention can never entirely contain the mysterious alien worlds of objects" (Bogost, 2012: 50). Indeed, Bogost goes to great lengths to outline the ways in which objects interact with the material processes of a camera and photographic processing to posit glimpses of how the camera – rather than the human eye – encounters and partakes in such alien worlds in ways that extend beyond but are still obliquely accessible to humans (after all, we made the cameras and view the pictures). Later in the chapter, in a process that combines listing with ontography, I view the process of creating/collating the plastic totem poles (and of documenting them through photographs and discussions) as a kind of 'exploded view' – but one which also underlines the synthetic and sticky qualities of plastics and that (re)connects with the specific social and geographical positioning of students at a particular school in Birmingham.

The final element that I wish to pick out of Bogost's remarkably rich theorisation of alien phenomenology is the notion of *metaphor*. Here, Bogost argues that although what I have called 'instrumental interfaces' might help us understand the physical components and processes that take place when a bat senses a fly, they do not enable us to describe how the bat *experiences* that fly, subjectively.

Thus – couched in the familiar notion of OOO that objects withdraw – 'alien' phenomenology acknowledges that any attempt to access the subjective experience of non-humans is always going to be somewhat flawed or partial. One response is that this is OK: it's good at least to try to understand the understanding of an-other, to empathise, as long as that move is (and this is tricky) not a colonial one or one that exercises unwieldy human power (e.g. Haraway, 1997; Taylor, 2019). Another – which aligns with Bennett's (2010) writing on anthromorphising – is to use the analogy or metaphor. Looping back to Harman's sense in which objects 'bathe' in one another, if anthropocentrism is ultimately unavoidable, the notion of the metaphor allows us to understand that, even if many qualities remain withdrawn, objects still do routinely encounter one another, with or without humans. As I indicate below, elements like titanium flow through the environment, interacting all the while. In practice, this means using metaphor (such as, "the bat . . . operates like a submarine") for "characterizing object perceptions", "rather than by describing the effects of such interactions on the objects" (Bogost, 2012: 64 & 67). Hence, we could also understand plastics not only *as* synthetic or sticky but *like* synthetic or sticky (or melting) – or other environmental elements as *like* plastics – in order to get some kind of a grasp on what it's like to *be* a prosthetic leg in a totem pole, or to *be* aluminium. Bogost returns to cameras and the ways in which they 'see' light levels and colours as a process analogous to (but not the same as) how humans see in different light conditions.

In this brief exposition of his work, I have outlined how – alongside other forms of (new) materialist and object-thinking, Bogost's alien phenomenology *might* offer further tools for thinking and doing, after childhood. I *experiment care-fully* with the notions of *instrumental interface, speculation* (again), *litany, ontography* and *metaphor* in what follows. But, I emphasise that Bogost's work *might* offer useful tools and my impulse to *care-ful experimentation*. Perhaps here I withdraw or diverge from the more radical elements of Bogost's work because, after all, I am still concerned with *children*, even if it may not always seem like it.

And it is here that, before turning to those experiments, I briefly refer back to critical race and queer theorisations of matter as a necessary second frame of reference in answering the question I posed earlier: of how to respond to plastics and other materials like them. Without putting too fine a point on things, recent ways of thinking about matter – and especially new materialisms – have been subject to some stinging critiques (see also Chapter 2). In my view some of the most important have been forwarded by queer and, especially, critical race theorists. To repeat my stance earlier in the book: in citing and seeking to take inspiration from this work I acknowledge my position as a privileged, white, heterosexual male scholar, and in turn acknowledge that that simple act of acknowledgement is a drop in the ocean in dealing with issues of race and racism both within and beyond the academy. However, although race is actually not such a prominent concern in this chapter as in the last (at St Paul's School), my interfaces with the different 'data' present in this chapter have nevertheless been patterned by the notions of *stickiness, toxicity* and *litany* that I outlined in Chapter 2. I therefore expand on my interpretation of those three terms throughout the chapter. Thus, in the rest of this chapter, I want to hold (new) materialist/OOO and critical race/ queer theories in tension – perhaps ironically, not in synthesis, being prepared for them to become *un*-stuck – in order to narrate some speculative encounters with and beyond plastics. If we remember that the (absent) bodies of which Huang speaks might not only be *raced* but *aged*, then we are in a position to consider how we might put in place some final ways of thinking and doing, after childhood.

The totem pole as ontography: starting again with plastics

A litany of plastic (Figure 8.1):

- a polyster onesie (featuring a design from Disney's film *Frozen*);
- five ceramic-coloured plants pots;
- the right leg from a mannequin (in pale pink);
- a vibrant yellow *Gatorade* bottle and a variety of other soft drinks bottles;
- a purple toy basket;
- some plastic-coated wires;
- a 'for sale' sign;
- a toy spade;
- some filing trays;
- indistinguishable coloured toy building blocks/cups/shapes;
- a 'bag for life'.

These – and literally hundreds of other – items formed our starting point for the totem poles exercise: the final workshop with 12–15 year-olds at the University of Birmingham School, who collaborated with us in the Plastic Childhoods project. Working with Birmingham-based artists, *General Public*, we wanted to trouble the students' relationships with plastic. After a full term of workshops and activities, the students had become familiar – almost comfortable – with plastics, knowing probably far more about different types of plastics and their environmental effects than did we adult 'experts'. From an environmental education perspective, we had succeeded. However, we wanted to afford a sense that, as Davis (2015) and Ghosh (2019) and others put it, plastics exceed human knowledge and, especially, they exceed the *horizons* of ordinary categorisation. Paralleling our work on 'other energies', we wanted to experiment with other ways of relating to plastics (see also the 'Living with Plastics' theme of the Climate Action Network for further examples of such experimentation: http://livingwithplastics.climate actionchildhood.net). If plastics can queer biological sex, sexuality and fertility, or function as hyperobjects at spatial and temporal scales both too small and large for humans to perceive, then their *phasing* into our programme of workshops could only ever have been partial, ephemeral, and modest (Morton, 2013). We needed to stick with, or, rather remind students of, the trouble: not of the trouble caused *by* plastics (about that they were well aware, and very militant) but of the very elusiveness of plastics and the other ways in which they could forge 'kinship' with chemical others beyond their by-now well-honed technical knowledges

Figure 8.1 A litany of plastic: a close-up of some of the items collected and set out before the totem pole workshop

Source: Author's photograph

about different types of plastics (after Haraway, 2016; see also the conclusion to Chapter 7).

The story of the totem poles is rhizomatic, arbitrary and synthetic. After Deleuze and Guattari (1988), it therefore starts midway – at least, in the lives of the plastics themselves. The litany with which this section starts offers a bare testimony to (or 'cut' through) the collection of objects in Figure 8.1, which in turn barely witnesses the full collection of objects in the room, and which in turn barely represents the arbitrary, almost random selection of those objects from the millions of tonnes of plastic produced, used and dumped each year. Whilst it would probably be a little disingenuous to claim that the plastics that ended up set out on the floor of the classroom at the beginning of the workshop 'selected us' as much as 'we selected them', there is nevertheless a sense in which their appearance as an assemblage – as *that* assemblage – was nonetheless partly out of our control.

A few weeks before the workshops, I met with the artists, Chris and Liz, to decide what kinds of objects we wanted. I was keen for a variety of kinds of objects – some had to be obviously relevant to 'childhood', but many should not be, and some should be the kinds of objects that prop up modern lives in this part of the United Kingdom (like clothes or medical products); cheapness was a factor; we also wanted different shapes, textures and colours, which might afford different sensory interactions and aesthetic judgments. Akin to Halberstam's (2005) notion of 'scavenging', Chris and Liz then set about searching through the material archives – we might even, after Chapter 6, term these 'archaeologies' – of plastics found in and around Birmingham. Some items were found in skips on the street. Others were bought at charity shops. Others came from recycling bins or were found in the home. Again, the *specific* items found us, to an extent: Chris and Liz did not select *that* mannequin leg above all others (there was not *that* much choice, locally) but selected it *because* it was a mannequin leg that presented itself to them; it mattered that it was *a* mannequin leg whose presence in the room would inevitably lead to a range of (potentially humorous) interactions, but it could, in effect, have been *any old mannequin leg*, because it was one of literally millions produced at the same time. So many of the items present in the school just happened to be in the right place at the right time (or, perhaps, the wrong time, because each received rather violent treatment from a drill to produce a hole so that they could be skewered onto a metal pole in making the totems). Whilst we can't quite say that the items 'selected us', then – there was a good deal of deliberation involved – there is a sense in which the *specific* items did, simply by being *that* pot in *that* skip in *that street* in Birmingham on *that day* that we were looking for plastics. Each item, like the doll in the apartment in Luz, and like the toys found in archaeological 'cuts', had a biography, in which their selection could only come midway through a lifetime lasting hundreds or thousands of years (even if, for many of their later years, as broken down fibres or microbeads circulating in oceans or the guts of fish, not resembling the objects with which we interfaced). Each item had been manufactured, transported, marketed, transported some more, bought, transported a bit more, used (probably transported during that time), and discarded (and transported yet again). Each item had a trajectory as well as a biography, with their presence on the floor of the classroom a temporary resting point – a temporary reinscription of value, of sorts – before moving on. Indeed, this was somewhat ironic, because many items had probably been discarded precisely in order to be forgotten (Huang, 2017).

The litany above, then, is an attempt to witness the slightly arbitrary, out-of-control sense in which *our* synthesis of objects, arranged onto the classroom floor, came about, midway through the lives of an unexpected cohort of plastics. From there, in groups of three or four, the students were asked to create totem poles of varying kinds, by slotting the objects using the pre-drilled holes onto flexible metal rods. They were given different instructions: make a totem pole (no specific parameters); make a totem pole that differentiates the usefulness of different plastics; make a totem pole that shows the value of different plastics; and so on. Figures 8.2 (no specific parameters) and 8.3 (value of different plastics) provide examples of the students' sculptures.

I would like to think of the totem poles – however tenuously – as a *synthesis* of litany and ontography. Individually and as a collection, they represented attempts to list – not so much through words, but through haptic and aesthetic judgments – and to put together a kind of 'exploded view' diagram of how plastics *might* go together. This is the "meanwhile" of plastics' biographies, as plastic "units reveal themselves" through the kinds of interfaces with humans and metal rods for which they were never designed (Bryant, 2014: 51). Indeed, by compromising their physical integrity or objecthood – by drilling a hole and impaling them – a little something of 'what lies beneath' is revealed, just like the damaged toys and bones found in archaeological or digital 'cuts'.

Notable, too, was a bizarre kind of *metaphorism* in students' arrangements of objects. Several of the totems in the first task – without parameters – ended up looking vaguely human-like, even though, aside from a few articles of clothing and a couple of helmets, few of the objects resembled or were designed to adorn human bodies (Figure 8.2). Interestingly, these forms were emergent. One of the researchers working with a group of 14-year-old British Asian boys recorded in her fieldnotes:

> In the initial exercise in which the team had to pick items without instructions the boys first seemed to go for objects which they recognised and were amusing to them (the 'trap phone' being a perfect example of this) or that they used themselves (e.g. 'ah yeh we've got to have the body wash, can't go without the body wash'). They then seemed drawn to the more colourful plastics (the elephant watering can, the fluorescent cup, the clothes), whilst also making sure they had a variety of different forms, texture and shapes in the mix.

Indeed, humour and decisions about form, texture and shape – both visual and textural – were a part of the entire workshop but especially this first 'free-form' exercise. But it was from judgments about which plastics *worked* well together – which sat well, which *stuck* onto the pole and *interfaced* in ways that were satisfactory – that two of the four groups developed more humanoid forms in which plastics became anthropomorphised (Bennett, 2010). The emergent

Figure 8.2 Totem pole created 'freeform' – with no specific parameters

Source: Credit: Jaskirt Boora

properties of the totem poles are to an extent evident in Figure 8.2: at the bottom, the trays and plastic packaging do not really form legs; but by the time the fake lawn was hanging from the pole (middle left of sculpture) the students felt that this looked like 'long hair' and so set about finding some tubing for the arms and the helmet for the head.

In the spirit of the 'pull focus', the workshops worked fairly well in both moving away from *and* returning to the (perhaps more conventional) forms of knowledges and interactions with plastics the students had had so far. The activity around the different value of plastics – which took place after the 'freeform' task – was particularly illuminating. Working with the same group of boys, above, the researcher noted:

> *The boys very quickly started to pick up objects from the pile and asses their value in terms of how often we use them everyday and how vital they are to our survival and wellbeing. The phrase they began to use was 'you can't have teeth without toothpaste and a toothbrush', 'you can't see without glasses', 'you can't have food without a tray', 'you can't have water without a bottle', 'you can't live without medicine', 'you can't have a baby without a doll', 'you won't have a head without a hardhat'. When constructing the totem they first started with a bunch of 'useless' stuff at the bottom which they felt also made a good base – the polystyrene block again, some bubble wrap and then the more vital stuff right at the top which as the doll, the glasses, the toothpaste, some medicine, clothes and again the phone. The totem was named 'Anabelle' after the possessed doll Annabelle from the horror film series – invoked by the toy doll at the top of the totem.*

This totem pole (not pictured here) was, then, formed through a curious mixture of judgments: about the material affordances of the 'bunch of useless stuff at the bottom' that nevertheless made the pole structurally sound; about the plastics that were, apparently, indispensable to everyday lives in a place like Birmingham; and about the decision to introduce a humorous popular-cultural reference to a horror film in a totem pole otherwise focused upon use-value. A highlight for me of their *talk* is how *they* decentred humans (including children), literally putting plastic objects before people through apparently twisted causal logics like 'you can't have a head without a hard hat' or 'you can't have a baby without a doll'. Knowing this group of boys, there was a dose of humour in these statements, too; yet, perhaps, in this humour lies a knowing irony about our reliance on plastics. They spoke of how "we are (en)plasticized", bound by a "plastic contract" (Ghosh, 2019: 277): of human lives *synthesised* and *stuck* with plastics.

Figure 8.3, by contrast, reveals a different set of logics around value. It may or may not be relevant, but this group was comprised of a mixture of boys and girls, and their focus was upon categorising the different value of different plastic objects. I worked with this group as they built this totem pole and recall in my notes:

> *This time we focused on value. Since the students already had a load of stuff in a pile from the previous task, they decided to work from that rather than get*

Figure 8.3 Totem pole created to reflect different 'values' of plastic

Source: Credit: Jaskirt Boora

some new items. They decided to use the pole to show different kinds of value. As they built, the totem pole kept changing as they realised the different ways they could categorise objects. 'The helmet could keep you safe'; 'yeah, but it could also keep you dry in a rainstorm'. 'The roses look pretty'; 'but some people put them on graves'. Once they had finished they then re-presented their ideas to the rest of the group. By this time they had a fairly firm and agreed stance on the value of the different objects. The helmet was for keeping people safe. The watering can was for keeping (other) things alive. The roses had emotional and social value – although they looked 'plastic' 'people put them on relatives' graves if they can't visit regularly enough to put real flowers'.

The totem poles and the chatter and activity around them represented, to me, vital snippets into other ways of living-with plastics: *speculative* senses of ephemeral, humorous, ironic, but knowing kinship with this stuff that, especially in the year the workshops took place, was becoming so reviled in British everyday lives. In the case of the totem pole in Figure 8.3, and talk of the emotional value of some plastic items, there was also a sense of the ways in which plastics, like sands, might also (help humans to) remember (Agard-Jones, 2012). Whether figuratively or literally, plastics bear scratches, dents, melted ages, twists of hands or twists of fate that led them to be present with us and present-to-hand in the classroom on that day.

I finish this section with two observations prompted mainly by critical race scholarship on the environment and environmental justice. The first – in the spirit of juxtaposition elaborated in the previous chapter – is to raise an open question about humour, silliness, speculation and 'voice'. In fact, this question is less about *race* only than about *intersecting* contexts and identities (Konstantoni and Emejulu, 2017). The question is: why are the speculative acts of a group of British Asian and Black boys at St Paul's Secondary school – such as destroying a wooden block house – viewed as 'disruptive' whereas those of a group of British Asian boys making a totem pole are viewed as 'ironic'? The obvious and simple answer is the institutional context and the expectations set around these two groups of boys (their being in two rather different secondary schools); yet the more complex 'answer' – which requires far more thought and deliberation – lies in a far more critical analysis of who gets to speak and act (and who is rendered silent and desubjectified) in environmental discourse (Jackson, 2015).

The second observation is an acknowledgement of the potential epistemological violence writ by learning-with (and playing-at) *totem poles*, and in using *plastics* – and *discarded plastics, of all things* – to construct them. Thinking from a post-colonial perspective, this is a hugely problematic, and colonising, and culturally imperial, way of acting. Whilst these may not completely assuage these effects (or my ongoing discomfort), this does not however mean that this task was morally or politically vacuous. One of the reasons for creating totem poles was to encourage the students to step out of their comfort zones and out of their immediate, every-day contexts – to consider how others might relate to plastic. Thus, we started the session with a detailed overview and discussion about what totem poles were, their histories, and their cultural significance. Another reason for creating totem poles was to provoke the students: to use material objects ostensibly classified as 'waste', as 'use-less' and as 'unappealing' to create a version of something that, in other times and places, has enormous meaning and value. But a final reason was to work *with* the ways in which totem poles are in fact conceived by the indigenous groups of the Pacific Northwest: the word totem pole in Algonquin means 'kinship group', and nods to our attempts to encourage the students to explore forms of kinship-with-plastics; totems may symbolise ancestors or cultural events, and, even if perhaps only intimated in passing and through humour, the flowers and the doll ('Annabelle') offered an echo of such forms of iconography; and, amongst many other functions, totem poles may be used to publicly ridicule someone, and so resonate with the senses of humour and irony that pervaded our workshop. Like plastics, the totem pole offered a *starting point*; but that does mean that either issue – the status of plastics, or the status of totem poles in our activities – is completely settled.

Beyond plastics: writing from aluminium, titanium and zirconium

I am aluminium: a litany (after Bogost, 2012: 18–19).

> *I am aluminium.*
> *I am aluminium. I am found in between 2.000 and 8.993 parts per billion (ppb) in the water that some young humans took from the taps in their home and put in some plastic tubes. In these concentrations, I am not dangerous.*
> *I am aluminium. I am made of very small particles. I am released into the air by smelting and by coal-fired power stations. There are lots of power stations in England. The nearest operational one is 45 miles away. I can hang around in the air for many days. I can be washed to the ground by rain, where I seep into the soil and the water.*
> *I am aluminium. I am used in my nanoscopic form in anti-perspirants and deodorants, where I am used to prevent toxins being released from human bodies, which might smell different without me. I can be absorbed through the skin. Some humans classify me as a toxin; others think I clog up the lymph nodes and cause certain kinds of cancers, although other humans called scientists don't have any real evidence for this. I am used by many of the young humans who are digging me out of the ground and finding me in tapwater.*
> *I am aluminium. I pass through the bodies of the young humans who dug me out of the soil, who breathed me out, who drank me in, who passed me out in their urine. Philip is the name of one of those bodies. He is 12 years old, and white. He found 437.214 ppb of me in the soil in his front garden. In the water in his kitchen sink, he found 6.038 ppb. When he breathed me out, there were 770.437 ppb of me in his breath condensate. There were 473.847 ppn in his urine.*

> *I am aluminium. I am classified by the United Kingdom's Air Quality Expert Group as a primary component of particulate matter, found in coarse dusts from quarrying, construction and demolition work and wind-driven dusts, alongside silicon, iron and calcium (https://uk-air.defra.gov.uk › documents › reports › aqeg › pm-summary). There is lots of building going on in the centre of Birmingham, a few miles from here.*
>
> *I am aluminium. I am found with and sometimes stick with lots of other stuff, especially after taking part in construction or demolition. "Construction and demolition waste (CDW) is one of the heaviest and most voluminous waste streams generated in the EU. It accounts for approximately 25% – 30% of all waste generated in the EU and consists of numerous materials, including concrete, bricks, gypsum, wood, glass, **metals, plastic**, solvents, asbestos and excavated soil, many of which can be recycled." (https:// ec.europa.eu/environment/waste/construction_demolition.htm)*
>
> *I am aluminium. I am found in much higher concentrations in the soil near major roads. I may not come from cars' exhausts, but cars blow or wash me to the sides of roads where I take a rest in the soft soil. Philip found 1186.861 ppb of me in the soil in the car park at the front of the school, by the main road. He told a slightly bigger human, who calls himself a researcher, that he walks along this road twice every day, on his way to and from school.*
>
> *I am aluminium.*

As outlined in Chapter 1, during the Plastic Childhoods project, alongside the workshops (above) and the app and interviews (below), we trained and worked with 13 students at the school to take samples of tapwater, soils, breath and urine. We also worked with environmental nanoscientists in my school to collect, process and analyse the samples they had taken, spending time to discuss the results afterwards. As well as plastics, we tested the samples for 27 other elements, including aluminium. Two things stood out. Firstly, we did not find any plastics in the urine and exhaled breath condensate (EBC). This was a relief – the presence of plastics, especially in students' urine, would have indicated some serious illness or defect in kidney function. Plastics were, however, present (in fact being visible in macro form as crisp packets, sweet wrappers and manner of other rubbish) in the soils, and it has been well established that there are microplastics in tapwaters (see above). Once again, then, we were *starting-with* plastics, which, as the litany of aluminium above alludes, were found alongside a whole range of other, just as interesting, and potentially just as troubling elements (I shall come to two others shortly).

Secondly, my conversations with environmental nanoscientists involved a degree of *speculation*. This is by no means a criticism or a fault with our methodology. The sampling techniques and analyses were very precise and rigorous when it came to determining quantities of each element in ppb. In our detailed conversations about the sources of the different elements, it was also possible to be precise about the *kinds* of places and processes from which each came: there is plenty of evidence that aluminium comes from coal-fired power stations, quarrying, construction and demolition work (alongside plastics and other earth elements), and that it is blown and otherwise transported through the environment. Yet one nano- or micro-particle of aluminium looks very much like another. It was, therefore, not possible to conclusively say where *our* aluminium, in our samples, had come from. Probably from local and regional power stations, building work and quarries: but which? And from when? Could the particles have been released during smelting works in the nineteenth century, when Birmingham was a hugely important industrial centre (built on *bricks*), being part of the infra-generational lives of more-than-human lives lived since then – inhaled, exhaled, ingested, urinated by several generations of residents? Thus, a measure of *speculation* as to the sources of the aluminium and other elements found in our samples was required.

And, thus, my use at the beginning of this section of a Bogost-inspired litany: an experimental piece of writing that I would hesitate to term 'poetic', although I nod here to the wonderful work of Clare Madge (2018), who uses poetry autobiographically to explore what she terms 'living on' from breast cancer. Whilst I am not claiming in any way to have produced a piece of poetry, the experimental, autobiographical litany above (on behalf of the multiplicitous 'I' of millions of aluminium particles) uses the speculative and the repetitious to afford a glimpse of the ways in which aluminium took part in what I earlier termed the 'instrumental interfaces' of the Plastic Childhoods project. Even if it appears somewhat ridiculous to write from the point of view of aluminium, the *effect* created, and, I hope, the *affect*, is to work *with* scientists (as with artists) to decentre human (children) within an account of the presences and circulations of the lives of metals (after Bennett, 2010). But, critically, this is not an effort to dismiss or move *beyond* childhood (Spyrou, 2017) but to experiment with new and perhaps strange forms of knowledge production in which children are implicated (which they are, very much, in the litany above, albeit from an estranged vantage point) (Prout, 2005; Aitken, 2010).

What I am calling for, then, is a kind of creative, perhaps 'radically' interdisciplinary (more-than-)childhood studies that admits rather than effaces the potential contributions of scientists – the latter being something that childhood studies scholars, including myself, have been doing for far too long. The conversation with scientists and with instrumental interfaces is not in the search for objective explanatory power but something else: new collaborative toolkits, possibilities and knowledges that foster new forms of experimentation and speculation as much as 'hard data' about aluminium concentrations (although those, too, I for one find interesting). Perhaps these forms of interdisciplinarity might only be 'radical' in the sense that they involve collaboration with disciplines further outside the normal round of interdisciplinary childhood studies – between sociologists, geographers, anthropologists, educators, psychologists (even) – to admit nanoscientists and artists. But my interest really is not in defining modes of interdisciplinarity, rather in the forms of *synthesis* that conversations like those witnessed in this chapter might bring. These forms of synthesis might enable us to see what *sticks together* in the methodological and analytical scavenging that has gone on between geographers, teachers, students, nanoscientists, artists, photographers and a range of others, each complicit at particular moments in the production of the stuff of which this chapter is composed.

Importantly, as I argued in the previous chapter, in these discussions, we (childhood studies scholars) must not lose sight of what these 'new' data might tell us, and what we might do next. From the litany of aluminium, a particular highlight, for me, is the set of questions that the data raise about air pollution. Birmingham experiences fairly high levels of air pollution (of NO_2 and other particulates) and, at the time of writing, was in the middle of an initiative to institute a clean air zone near the city centre, which would exclude the most polluting cars (www.birmingham. gov.uk/info/20076/pollution/1763/a_clean_air_zone_for_birmingham). Yet whilst the levels of aluminium (not produced by cars) found in the soil (especially) by Philip are not necessarily at levels dangerous to human health, to take the perspective of aluminium is to ask again – just as some environmental commentators have done with plastics – why certain kinds of environmental issues, practices and toxins come to dominate. This is not a critique of the Birmingham Clean Air Zone per se, and nor is it a rebuttal of policies that appear anti-car – that is not my politics. Rather, as with the questions raised by critical race scholars, it is a nudge to consider the manifold complexities and, potentially, slow forms of toxic violence (Agard-Jones, 2013; Davies, 2019) being writ upon environments and children as they grow up with aluminium and other elements as kin – and as levels gradually increase over time. What if, in the future, those levels of aluminium *do* become potentially harmful, in particular places? What if those places (like the area around the school) are not as high profile as the city centre, or are characterised by levels of social and/or economic deprivation?

A related consideration has very much to do with the *source* of aluminium in the environment. Whilst we might speculate, it is very likely that at least some of the aluminium in our samples came from quarrying, construction and demolition sites, probably not that far from the school. Thus, our data provide another perspective – literally – on recent literatures that have tried and often struggled to articulate children's entanglements with the Anthropocene and especially with its *geological* stratification. The 'cuts' examined in Chapter 3 are a case in point of this struggle – they are metaphorical as much as literal 'cuts' through the earth. The fabulous work of recent childhood scholars like Gallagher (2019; see Chapter 4) and Hadfield-Hill and Zara (2019, on geological subjects in India) gets us closer but relies largely on media discourses and children's testimonies. As I have repeatedly argued, there is nothing *wrong* with either – and I do not purport to have found the solution – but I do think that collaborations with environmental nanoscientists and others, which enable us to witness some of the material presences of geologically derived elements moving into, past, through and out of children's lives, might be a step along the way. Critically, these observations also prompt important, if not vital, questions for future research, policy-making and practice: what is the scope for a larger-scale, more longitudinal, study of exposure to elements like aluminium in a place like Birmingham? What are the effects of exposures not just to singular but multiple toxins, do these vary with geographical or social difference, and do choices of which forms of environmental pollution to address reflect forms of silencing or othering (compare Jackson, 2015)? What scope is there for more sustained studies of children who live near or – in some contexts – work on quarries, building or demolition sites, beyond the fascination with interactions between children and *plastics* at sites like rubbish dumps (compare Katz, 2018; Brickell et al., 2019)?

Our collaboration with students, artists and photographers has provided another way to speculate on from our elemental data: namely, to articulate and visualise encounters with plastics, metals and other elements in ways beyond the litany (although which retain something of the litanic style). I come to our work with students in the final section, and have already discussed the totem poles, so I focus here on ways I have worked with artists, photographers and designers to visualise the ontographies of two other elements: titanium and zirconium (Figures 8.4 and 8.5).

In some ways, I want these ontographies to speak for themselves. In collaboration with the artists and graphic designer, they have been carefully visualised to offer a kind of 'exploded view diagram' for each element that *stylistically* – perhaps *metaphorically* – maps out the interfaces through which each element appeared in our samples. Specifically, they provide another way to visualise instrumental interfaces (how the two elements 'bathed' in the FTIR and ICPMS machines), via two rounds of (speculative) conversation and collaboration: with environmental nanoscientists and the artists/designers. As I argue throughout this chapter, and in others, Figures 8.4 and 8.5 are examples of what 'radical' forms of interdisciplinary collaboration might *look* like, in thinking and doing, after childhood. They are also examples of what those forms of narration and visualisation might look like, in thinking and doing, after childhood – to be added, for instance, to the disjointed 'silliness' of Chapter 7, or the litanies in that chapter and this. But they are partial, and limited, too, and require a sense of context. For instance, Figure 8.4 would perhaps be understood in a different light if it were read in the context of moves in some countries (such as France) to have titanium classified as a carcinogen. I ask – speculatively – what that might mean for further investigations into levels of titanium found flowing into, through and out of children's bodies and environments.

Plastic knowledges, plastic experiences: putting children and young people (back) into the mix

A central element of the Plastic Childhoods project was an app – also called Plastic Childhoods – that enabled students to take photographs of plastics in their everyday lives and to comment on them under three broad categories: food, water and leisure. After a week of using the app, we interviewed each student individually for around an hour, laying out photographs taken within the app onto a large sheet of paper. Students selected photographs and talked us through their significance; we encouraged them to divide, group and make connections between photographs according to their own thoughts and classifications. Thus, some divided plastics into how commonly they encountered them, others into their sensory interactions (whether they saw, touched, smelled them), and others into their actual uses

Titanium

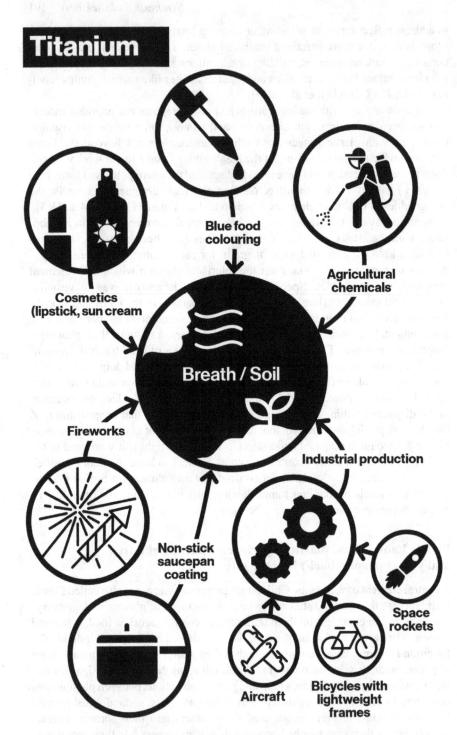

Figure 8.4 An onto-cartography of titanium, inspired by Bogost (2012: 51).

Source: General Public/Keith Dodds

Zirconium

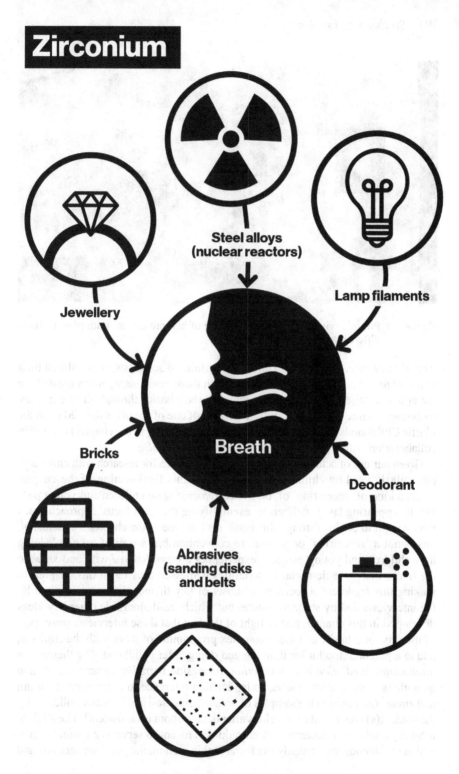

Figure 8.5 An onto-cartography of zirconium, inspired by Bogost (2012: 51).

Source: General Public/Keith Dodds

Figure 8.6 Philip's 'visual web' showing some of his interactions with plastics (note different 'uses')

(see, for instance, Figure 8.6). During the ensuing discussions, we explored their knowledge of plastics, their interactions with them, and – using maps created via the app as a baseline – *where* they encountered plastics through their everyday mobilities. Figure 8.7 offers a stylised version of one of the maps derived from the Plastic Childhoods app (for Philip, who appeared earlier in the chapter) created in collaboration with a graphic designer and *General Public.*

Given our use of actually fairly 'traditional' methods for research with children – particularly used by children's geographers – this final section of the chapter offers a kind of 'recentring' of, or, better, *refocusing* on children and young people. Its appearing last is deliberate, exemplifying the 'pull focus' approach that I have sought to deploy through this book. As I outline in the chapter's conclusion, this is not a 'recentring' or 'retreat' to convention but a way of acknowledging how children and young people's experiences are *synthesised* with and *stuck* to the (more-than-)plastic social-materialities witnessed thus far in this chapter. In making this argument, I focus on a subset of key themes that were inspired by the interviews and by students' voices but which, read alongside other key ideas developed in this chapter, and in light of the fact that these interviews were 'part of the mix' of a larger and more complex programme of work with the students, add to a potential toolkit for thinking and doing, after childhood. The themes are *habituation, mediation* and *interaction.* Alongside narrative summaries of, and quotations from, the interviews, I also include reference to aluminium, titanium and zirconium found in the samples of breath, urine, soil and/or water collected by the student(s) concerned (in brackets after details about each student). The slightly arbitrary choice of elemental concentrations is meant to serve as a reminder that, whilst childhoods are certainly the focus in this section, the previous sections and

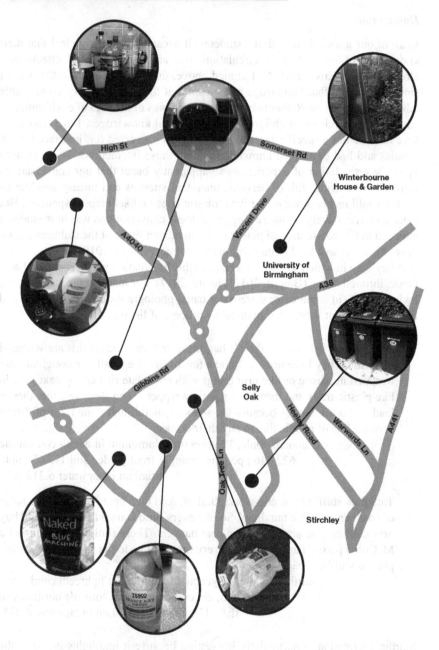

Map:
Overview of collection area

High St

Somerset Rd

Winterbourne
House & Garden

Vincent Drive

A4040

University of
Birmingham

A38

Gibbins Rd

Selly
Oak

Heeley Road

Warwards Ln

A441

Oak Tree Ln

Stirchley

Figure 8.7 Stylised map showing how Philip's everyday mobilities led to interactions with plastics

Source: Credit: General Public/Keith Dodds

chapters have demonstrated that, now we are aware of the presence of aluminium (for instance), we should not instantly forget it: we must *also* try to retain our focus elsewhere, whether as synthesis or juxtaposition (Chapter 7).

Habituation

In all of our interactions with the students, it became abundantly clear that their knowledge of plastics, their circulation, use and environmental effects was already very sophisticated. As I argued above, their participation in the project simply bolstered that knowledge and meant that they became very comfortable talking about plastics. Rather than repeat the findings of decades of environmental education research about children's environmental knowledges, I therefore seek here to tease out a specific facet of how students interacted (or, interfaced) with plastics and had developed knowledges about those interactions. These centred upon the articulation of experiences so apparently banal that our conversations were often faltering, full of nervous, uncertain silences and uneasy laughter as students told us about recycling bins, plastic cups and sellotape dispensers. But what was, recursively, so interesting about these conversations was how students attended to the *habituation* of plastics with/in human lives: of the stubborn *stickiness* of plastics as a kind of contemporary contract (Ghosh, 2019).

A key way in which students attended to this habituation was stylistic and aesthetic: through *litany* (Bogost, 2012; Huang, 2017). For many of the students, the laying out of (in some cases dozens) of banal photographs of plastic items, and then talking about them, was an uncanny process of listing.

> Urm, I think the school, there's baguettes and sandwiches that are wrapped in plastic. They have urm . . . quite a few of like the small like doughnuts are wrapped in plastic or like urm things with chocolate rice crispy cakes are in like plastic trays and then have plastic wrapper round them and the, there's load of plastic bottles because the school gets through quite a lot of drinks because lots of people like to buy their drinks.
>
> (Leonard, male, 12 years old; aluminium in breath condensate 622.446 ppb; zirconium in front garden soil 115.650 ppb; titanium in tapwater 6.312 ppb)

> Toiletries stuff, music and work, kind of. And then here, we have like loads of coat hangers, like hundreds. So like every time you go to the shop and buy new clothes, you always get the coat hangers. These (holding a picture of a McColl's plastic bag) obviously every time you go to the shops, they always put our stuff in a bag.
>
> (Martha, female, 13 years old; aluminium in breath condensate 918.316 ppb; zirconium in garden 'outside our block of flats' 134.645 ppb; titanium in tapwater 0.035)

Martha's quotation is particularly interesting because it highlights not only the (rather obvious) ineffability of our habituation to plastics ('stuff'/'kind of') but the sheer quantity of that stuff: of her family's having 'like hundreds' of coat hangers (other students talked of having 'hundreds' or 'thousands' or just 'ridiculous amounts' of plastic objects from plastic bags to pens). Thus, a function of our habituation is omnipresence: there is a certain irony or humour to Martha's knowing articulation of having 'hundreds' of coat hangers, with the unsaid (but again rather obvious) question left hanging: what would anyone do with such a quantity of coat hangers? The relative ineffability (and banality) of plastics, and their sheer quantity, were accompanied by observations by students about how *frequently* they used or came across plastic items (Figure 8.8 being Maura's).

Beyond listing, students' articulations of their habituated interactions with plastics exemplified, once again, the idea of the 'pull focus'. The key argument here is that, *at the same time* as attending to the pull focus when it comes to *childhood*, it is important to attend to the pull focus for those matters with/in which childhoods are 'entangled' (or, synthesised). In other words, there are multiple pull foci, occurring simultaneously, as machinic assemblages (plastics-childhoods) operate dynamically (Bryant, 2014). For instance, many of the students spent time talking about the products that they were using that were contained *within* plastics – the taste of food, the use of a particular shampoo or moisturiser because it dealt with a dry scalp or area of skin. Others questioned whether, when they used a product packaged in plastics, but that product were not itself plastic, they were 'using' plastics or not:

"I use those everyday . . . this I use weekly, it depends on how my hair is [giggles]"

(Laura, female, 13 years old; aluminium in breath condensate 770.436 ppb; zirconium in garden soil 116.317 ppb; titanium in downstairs toilet tapwater 519.150 ppb)

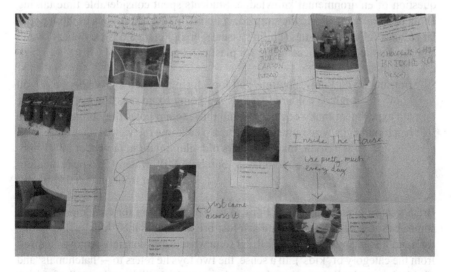

Figure 8.8 A detail from Maura's visual web, categorising the frequency with which she interacted with particular plastics, and the sensory registers (sight, touch, etc.) through which she encountered them

"Okay, it depends though . . . do I use the plastic wrapper or do I use the kitchen roll?"

(Basirat, male, 12 years old; aluminium in kitchen water 5.373 ppb; zirconium in back garden soil 136.610; titanium in kitchen water 1012.572)

Basirat's question has a twofold implication: on the one hand, he asks whether our habituation to plastic is so complete that we may not even see its presence as a wrapper as our 'use' thereof; on the other hand, he is asking about plastic's relationship with and synthesis within material cultures involving other stuff (like kitchen paper). Meanwhile, Laura's giggle when talking about her hair was part embarrassment but partly symptomatic of the observation made above: a nervous outburst at the sheer banality and ineffability of plastics – and our habituation to them.

Mediation

The example of Basirat's kitchen paper offers an entry point into the second and third themes that emerged from conversations with students: mediation and interaction. In terms of plastic *mediation*, here I follow both Bryant's (2014) conception of media-machines and my reading (in this chapter) of plastic's capacities to mediate like water (Neimanis, 2017) and sand (Agard-Jones, 2013). Specifically, the students highlighted two kinds of concern, both of which relate to plastic's *sticky* properties: knowledge of the material cultures with/in which plastics were firmly entrenched and anxieties about the material constitution of those cultures *as* plastics.

Although ostensibly similar, the first point relates (again) to the more familiar question of environmental knowledge. Students spent considerable time talking about popular cultures and the simple fact that so many expressions of those popular cultures relied on plastics (Philip's visual web in Figure 8.6 is exemplary – and, again, a litanic form of accounting for plastics). For instance:

But that's you know, anytime you look for plastic it's everywhere and you know it's brightly coloured so kids are going to go nuts for it. Like during the experiment that week that we did those totem poles they brought out the hatchimal eggs, you remember ferbies. You know the weird. . . .

(Brian, male, 13 years old; aluminium in tap water 5.001 ppb; zirconium in 'Sarehole Mill' water 0.276 ppb; titanium in front garden 0.849 ppb)

Brian's sense is that plastic functions so well as a medium for contemporary consumer cultures because of what it affords: bright colours that 'kids' will go 'nuts for' (and it is interesting here that Brian somewhat dispassionately removes himself from the category of 'kids'). In a sense, the two toys he refers to – 'hatchimals' and 'ferbies' – require little explanation: not because they will be universally familiar, but because their identity does not really matter – here, hatchimals and ferbies could, in the sometimes arbitrary style of Latourian listing, equally be Tamagotchi or Barbies or Pokemon (Horton, 2012). By the way, Brian's samples were interesting because he took stream water and soil samples from a place called 'Sarehole Mill' – a place that he liked to visit with his family. Notably, concentrations of several elements, including zirconium but also boron (102.034 ppb), titanium (22.311 ppb) and arsenic (1.235 ppb) were far higher than in other samples and, in some cases, above the recommended safety thresholds. In the case of stream water this will be because the water was not treated; in the case of soil, the explanation is not so straightforward. Whether this equates to greater exposure in Brian's case would require further testing: but, as I argue in this chapter's conclusion, we would not even be able to raise this question had we not *also* asked the students to tell us about their everyday experiences and mobilities.

Returning to plastics: Brian's is a well-known (although humorously articulated) point about how many elements of popular cultures are made of plastic. This is a pretty obvious and uncontentious point that requires little further discussion. Yet Brian's point is also one about the material affordances of plastics – in this case, in terms of colour. Other students spoke at length about the haptic affordances of plastics: about the feel of a computer keyboard as the fingers tapped away; about the smoothness of books covered in plastic by a laminating machine to protect them from damage. Yet some of the students also expressed concerns about plastic's very material properties – almost a sense of its agency – and what this meant in terms of how plastics interfaced with humans and others. In doing so, they afford a sense of the 'other energies' – other embodied operations – with plastics (Philo et al., 2015; see Chapter 7). The example of microbeads – which were entangled with/in popular cultures contemporary to the research in 2019 – should demonstrate what I mean.

During one workshop, we asked the students to walk around the school collecting as many plastic items as they could find. Many of them found colourful small beads – a little like microbeads – all around the school. Indeed, Maura (in Figure 8.8) took a photograph of these in the school whilst using the Plastic Childhoods app, saying "[t]hey are small little balls of plastic that when they go into water they expand but I am not sure why they are at school". As Maura said during the interview: "we [the group] found them *everywhere*". Brian usefully (and indirectly) answered Maura's question about why they were at school, explaining that the balls are called 'orbies' and were a popular toy meant to help with stress: "Yeh all of those orbies, everywhere. I get they're used for like soothing stuff and people like to be calm but you know, one – why is that a toy and two – why does it need to exist?" Another student, Georgia (a 15 year-old female), pointed out that she had seen the beads on *YouTube* but that "they used to be this kind of huge trend" but were now less popular.

These conversations witness something of plastic's many capacities to act as a medium: to *synthesise* social-material concerns and mediations. There is not space to discuss these concerns fully here, especially because some are somewhat speculative and/or implied rather than explicit. Yet (in the now-familiar style of listing), they synthesise: the banal, everyday ways in which popular-cultural trends come and go (Horton, 2010), and in which plastic 'crap' (Huang, 2017) is a key constituent; the symbolism of these beads as a slightly larger version of the microbeads – partly decomposed and found widely in waterways and the stomachs of fish – that have been a key focus in debates about the dangers of plastics; a textual ontography (Bryant, 2014) of plastics as they (literally) 'bathe' in water and, at the interface, expand and turn into a jelly like substance; and, at a stretch, a nod (in Brian's comments about stress) the rise of anxiety, stress and other mental health issues amongst (especially) teenagers in Britain, and at responses varying from digital media resources (Goodyear and Armour, 2018; Chapter 5) to the creation of tactile plastic beads designed to reduce stress.

Interaction

Whilst many of the students recognised the usefulness of some plastics (and a look back at the discussion of the totem poles serves to underline this), this latter discussion about the 'orbies' speaks of the more insidious and, to use a term I am adapting from critical race and queer theorists, 'toxic' capacities of plastics. Critically, and again with a nod to the totem poles, whilst inferring that plastics were in part to blame, many of the students just *started with* plastics. They thought, instead, about *interaction* as a form of synthesis: about how, if we are 'stuck' with plastics, we forge queer kinds of kinship in which they (as students), plastics, and other non-humans were involved (after Haraway, 2016).

In arguably the most simplistic and obvious way – obvious both from the point of view of environmental knowledge and the sheer size of the plastic items involved – several students told us about how they encountered 'litter' on their journeys to school, the shops or elsewhere: effectively, macroplastics (with individual pieces >5mm in size). Litter was often figured as a ubiquitous, amorphous 'mass' – redolent of the litter so frequently identified, by children, in studies of their experiences of (especially urban) places (Horton and Kraftl, 2018). 'Litter' is, in fact, a useful entry-point here. It serves as a reminder of the importance of (re)focusing on that central preoccupation of childhood studies – children's mobilities – because it was through reflection upon everyday journeys that students identified the problem of litter as well as a vast range of other plastics (see Figure 8.6 for just one example). But it also serves as a reminder of how – in constituting 'litter' – plastics are, as I have already argued, not the only problem. Starting with plastics helps: but plastics are synthesised, or entangled, or stuck, or matted (or whatever-ed) with other stuff to make 'litter', as the Luz doll, the vortex and other collections of stuff featuring throughout this book have attested.

It would be a stretch to figure litter as 'kin' – although some of the stuff with which we made totem poles would virtually qualify as litter. But, in telling us the story of her pet parrot, Hadassah, a 13-year-old girl, perhaps brings us closer to something of the queer kinds of multispecies and '(en)plasticised' (Ghosh, 2019) kin-making that queer scholars like Heather Davis (2015) are concerned with. Talking about a picture of her parrot's cage that she had taken using the app, the following conversation ensued:

HADASSAH: so basically, we started feeding him [parrot] out of like, out of plastic bags and everything, yeah and then recently – a few days after I took this picture he passed away

INTERVIEWER: oh no

HADASSAH: Um, so I'm not sure if like- because he's one of those parrots that he eats anything he sees

INTERVIEWER: okay. . . .

HADASSAH: So he *would* eat plastic if he could see it, we had to keep it away from him. And it's the same with my cat, she won't eat it, but she'll *play* with it and bite it a lot and I don't know how that will affect her and everything. Um, I just thought that it *might* have an impact because we put his food in plastic boxes and everything, so I just felt that like maybe it might have something, and maybe because of his death, maybe it happened because of the plastics and stuff

> (Hadassah, female, 13 years old; aluminium in top floor
> sink 4.652 ppb; zirconium in front soil 124.211 ppb;
> titanium in top floor sink 453.491 ppb)

Once again, there is a little *speculation* here about what might have caused Hadassah's pet parrot's death. However, as Chen (2011) reminds us, for instance, toxic substances like plastics are re-making categories and boundaries of life and death. For Hadassah, plastics offered a handy, cheap way to keep her pet parrot alive: the plastic-parrot synthesis was an important one in the making of multispecies kinship (Taylor, 2013). Indeed, that kinship might arguably extend *to* plastics: to Hadassah's living-with plastics *and* pet parrot. Yet, Hadassah considers, those very same plastics may have been a contributing factor in the parrot's death – alongside 'stuff' (as she ends the extract above with the phrase 'plastics and stuff').

Finally, if in Hadassah's experience this other (potentially toxic) stuff is not articulated, then other students – again reflecting on the plastic-popular-media cultures in which most British youth are stuck – grappled with plastic's synthesis with/in other forms of toxicity. For instance, Maura talked about the chemicals involved in a 'bath-bomb'-making kit she had received recently, but then spoke at greater length about her friend's fascination with making slime – another popular-cultural fashion in Britain at around the time of the research. She then reflected:

> Yeah, it's like online, and so, like, um, if you've been watching the news and everything, they've been saying what they've been adding into slime, so like micro beads.

She continued that there were two problems with this trend:

> And um, normally when people play with slime, they play with it once or twice, and then they throw it away and like that's a bit of a problem . . . And then like how you make slime, there are many ways, but the most popular way is um using borax . . . It's like a really dangerous chemical, it can burn your skin.

Aside from the references (again) to microbeads and the 'throwaway' nature of slime, the most striking element of Maura's description is the danger posed by the borax that is used to make slime. Although the many websites dedicated to providing instructions on making slime do not indicate that the chemical is 'really dangerous', they do, perhaps paradoxically, advise against prolonged skin contact and the use of protective gloves and eyewear when handling borax. Again, I would not make any claims about either gender or ethnicity here; yet, the work of critical race and queer scholars – and especially Chen's (2011) analysis of *Thomas the Tank Engine's* lead paint – is evoked in Maura's sense in which the touching of a toy poses potentially direct health risks to children. Critical here was starting-with plastics: both in getting us to the point in the conversation where she spoke about slime and in recognising plastic's synthesis with other (in this case potentially more toxic) chemicals in the making of slime. Indeed, the qualities of slime – *synthetic* and very much *sticky* – offer a rather fitting way to draw to a close my analyses of plastics in this chapter.

Conclusions

The example of the Burger King Meltdown that began this chapter acted as a useful pathway for expanding further the notion of the 'pull focus' as an analytical device for thinking and doing, after childhood. In particular, both childhoods *and* plastics – together and apart – move in and out of focus in this chapter. Childhoods and plastics are blurred and obscured, in different times and places, but sometimes simultaneously. In fact, they seem to interact with each other with particularly poignant, potent effects and affects, at least in the discursive idea(ls) implied by the 'Meltdown'. Thus, the idea of the pull focus applies not only to childhoods – and to children like the students at the school – but to the matters with which they interface (a little like the idea of 'biography' could be applied to objects as much as humans, as Chapter 6 showed). A key contribution of this chapter has therefore been conceptual and methodological: in particular, I have called for forms of interdisciplinarity that may or may not be 'radical' to all eyes but that, certainly, involve collaborations between a very broad range of fields that rarely come into conversation: geography, education, art and environmental nanoscience. To me, such forms of collaboration (although not necessarily in this specific configuration) are a vital constituent of thinking and doing, after childhood, because they allowed the genesis of a range of theoretical tools related to *synthesis* and *stickiness*.

The forms of instrumental interfaces I have witnessed in this chapter – the analyses of elements such as aluminium, zirconium and titanium – are hugely enriched by environmental nanoscientists with the equipment, skills and knowledge to quantify and distinguish between those elements, and to advise on their potential sources and impacts. Similarly, our interfaces with a range of found plastics (whoever found what or whatever found whom) represented, in Bogost's (2012) terms, forms of *litany*, *ontography* and *metaphor* – in themselves key methodological tools – that would simply not have been possible without the care, skill, creativity and resourcefulness of Liz and Chris at General Public Projects. In all, as I have argued, these forms of collaboration were less about being blinded by science and/or art, or seeing a 'solution' outside the social sciences, but in fostering conversations, happenings, material interactions and affects that effected arts of (not) noticing children and childhoods. Crucially, as in other chapters, these arts of (not) noticing were not only theoretical (not that there would be anything wrong with that). Rather, they drew attention to stuff that *matters* and which traditional social-scientific approaches to childhood cannot, in my view, fully apprehend: to our habituation to plastics and ways in which we can unsettle them through (for instance) making totem poles and to the many ways in which aluminium travels around, into, through and out of children's everyday lives and environments.

At the same time, however, I ended this chapter by deliberately returning to the experiences and voices of children and young people themselves. In theorising after childhood, this should be interpreted neither as a simple 'return' to questions of childhood nor as a retreat from the more radical implications of thinking and doing, after childhood. Rather – and this might just be semantic – I again would argue for a move away from notions of de-/re*centring* (Spyrou, 2017) towards notions of focus and (not) noticing. Whilst visual metaphors, these notions are not just about the *visibility* of childhoods and child*ren* in our analysis, although, as I argued in Chapter 5, questions of visibility have become ever-more important where children, young people and digital media are concerned. In contrast, notions of focus and noticing describe *differential* modes, registers and intensities in sensing children and childhoods – whether via the embodied acts of creating totem poles or the instrumental interfaces involved in collecting biological samples. In this book, childhoods are always in mind – I tried not to lose sight of them entirely – yet I have argued for an array of approaches that might enable us to sense childhoods differently. This might mean sharpening the focus on other matters (for instance, aluminium) whilst deliberately blurring or obscuring children, but that does not mean that children were either absent or decentred. Any notion of 'centring' implies a hierarchisation of human/non-human that is both at odds with the theoretical underpinnings of this book and with the kinds of synthesis and stickiness I have witnessed in this chapter. Furthermore, any notion of 'decentring' implies absence, and I have *also* not sought to render children or childhoods absent. Hence, *synthesis* implies the ways in which children and childhoods have always remained – however blurrily – part of the mix, whilst *stickiness* denotes the ways in which they could not and should not be detached from some of the key contemporary challenges I have sought to broach throughout this book. Methodologically speaking, this means that – as I have demonstrated – attending to some of the very fundamentals of childhood studies (such as students' everyday mobilities) remains as important as ever as part of this mix. However, as I will draw together in the book's conclusion, doing so might enable us to broach a range of pressing questions and concerns that, alone, those more conventional approaches cannot.

9 Conclusions
After childhood

This book has developed and exemplified what it could mean to think and do childhood studies, *after childhood*. It has set out a series of theoretical and methodological tools for bringing childhoods into and out of focus in ways that could enable new ways of looking at complex issues and challenges in which children are implicated. Whether this interdisciplinary endeavour might be termed 'childhood studies' or 'children's geographies' or something else – and whether in fact this really matters – cannot and perhaps should not be settled here. Indeed, if, as I explored in Chapter 2, one of the most pressing tasks when it comes to studying age is to engage in analyses of generational ordering, then the status of childhood as an object of study might shift just as much as it has throughout this book. However, those same analyses might also require a return to childhood/s as it/they is/are positioned within generational logics and discourses in particular contexts. In other words, a focus on a phenomenon such as generational ordering always brings with it the question – at least for those of us who are concerned with children: *where are the children, precisely*? This is one of the questions with which I began the book, and one that sits at the heart of my insistent commitment to the 'pull focus' and 'arts of (not) noticing'.

In turn, I argue that explications of generational ordering offer but one (albeit vital) way in which studies of age – and of childhood specifically – might proceed if they are to engage with some of the critical questions that Punch (2019) and others have raised about childhood studies as a field. Thus, I have in this book imagined what childhood studies might look like, *after* childhood. Generations have been one key touchstone, and in theorising *infra*-generations I have sought to push generation-thinking into new domains and into conversations with new disciplines (like archaeology), phenomena (like historical 'cuts') and theories-methods (like speculative fabulation). Yet the book is also made up of a range of concerns that go well beyond generation: from the (en)plasticisation of life (Ghosh, 2019), in which children are particularly implicated, to the visibility and circulation of childhoods, in and through digital technologies. Arguably, these kinds of concerns matter just as much as generations to the politics and pragmatics of being a child, in particular times and places, now and in the future.

Certainly, however, and recursively, the kinds of claims above raise further questions about what matters to children, childhoods and childhood studies – and how, and why, and who gets to make these claims. I am acutely aware that, in the spirit with which I began this book, my positioning as a white, male, privileged academic means that I both get the chance to stake out a sense of 'what matters' (in books like this) and that I am necessarily blinkered in my view. My hope is that, nonetheless, this book has attended to and developed a range of theoretical and empirical considerations usually excluded from childhood studies (albeit not necessarily intentionally, I think). These considerations will of course remain partial. Yet, empirically, in drawing attention to energetic phenomena, to resource power and to the intersections of such forces with young people positioned differently by geographical location, ethnicity, socio-economic class and behaviour, the book has attempted to grapple with issues that seem to matter not only to me, but to children, young people and the many adults who live with or advocate for them. Conceptually, the inclusion of a range of forms of material-thinking – particularly from object-oriented ontologies, critical race and queer theorists – has provided the inspiration and grounding for a suite of (again necessarily partial) tools and terms that might progress ways of thinking and doing, after childhood, which in turn could help broach those issues that matter.

There is also a tension between developing ways of thinking and doing that are intrinsically interesting, important or fun, and those that are deployed in the hope of wrestling with some bigger question or addressing this or that contemporary issue. I want to stay with this tension in structuring this brief concluding chapter. The chapter does not draw out a programme for future modes of thought or action or simply list off the many concepts and terms I have developed in this book. Doing either would run counter to how I understand what it means to work *after childhood*. Rather, I draw out three key strands from across the book, which might land in particular ways. These draw together and hold in tension the major conceptual, methodological and empirical contributions of this book, which are outlined somewhat more systematically in Chapter 1 and in conclusions to the individual chapters. The three key strands – which may not always have identified as such but are at least implicit – are *trauma*, *silliness* and *interface*.

Trauma is evident throughout this book. I point this out here not to sensationalise some of the examples but to point to how academic modes of analysis and reportage inevitably desensitise some of what it means to experience life in challenging circumstances. Despite all the clever word play, exposure to plastics – and to plasticisers – can cause very real trauma, especially for those workers who touch them in recycling plants every day, and especially for younger workers, who are generally more physiologically sensitive to their effects. Thinking back to how this book began: to live in the informal settlements of São Paulo or to be exposed to the long-term physical and mental health effects of air pollution – both are forms of intractably complex, 'slow violence' (Davies, 2019), with traumatic effects both acute and chronic. The 'cuts' into and across the earth that I have discussed throughout this book may be multiply traumatic: the damage caused to the earth itself by the production, divestment and burial of our massified detritus in sites as diverse as the eco-park in Guaratinguetá and archaeological 'cuts' in places like Cambridge and Shabona Grove; the markings of objects that may be symbolic of the loss of a child, or the trauma of experiencing socio-economic decline; the bones of children who succumbed to an early death as the result of intersecting social and environmental threats. Similarly, the experiences of some young people in Birmingham and Brazil are tied together by forms of stigmatisation and exclusion that can be traumatic: what must it mean to be marked out by attending a school for 'excluded pupils', or to have to constantly rely on neighbours to feed oneself and one's family?

As I argue in a moment, there have also been many 'lighter' moments in this book. But it is important to recognise that differential forms of trauma afflict children and young people around the world. Although thinking and doing, after childhood, will likely not directly solve any of those experiences, it does offer a set of conceptual languages for articulating how and why allowing children and childhoods to slip out of focus might enable a consideration of such forms of trauma to emerge all the more powerfully. Zooming out again, trauma also relates something of the conditions associated with social and environmental predicaments that seem – if the circulation of discourses on social media is anything to go by – to particularly affect children and young people. Indeed, questions about climate change and the mental health effects of digital media appear to be particularly profound for *this* generation of children and young people, now and into the future. The term 'trauma' seems to convey something of our collective earthly condition that, if no one else, the children and young people campaigning in #climatestrikes themselves identify.

Throughout the book, however, I have also attempted to introduce *silliness* as a necessary counterpart, corollary or accompaniment to trauma. The two are not dialectically opposed; indeed, I have deliberately chosen silliness because it sounds so frivolous, so ephemeral, so unimportant, so *childish*. But I could also have chosen other terms. Silliness is not meant here as a pejorative term. Nor is it meant as a rebuttal of or diversion away from 'robustness', although I have advocated some tentative forms of 'loosening' at various points in this book. Rather, I use silliness to loosely capture a series of modes of being, interacting, researching, writing and experimenting whose potential might – if not (only) reduced to the status of childish frivolity – offer generative insights that could work against or mitigate forms of trauma. I could just as easily have used the term 'speculation', but some of the methods and forms of narration I advocate extend beyond Haraway's (2011) use of the term. Thus, on the one hand (and to repeat, I do not mean this as a way to belittle either Haraway's work or the forms of expression I witnessed), speculative fabulation may offer a particularly apposite way to work across human-non-human generations in forming queer or other-worldly kinship relations. Throughout the latter parts of the book, I tried to draw attention to ways in which children and young people were implicated in forms of imagination, play or just *being* that were speculative: from a playful and apparently disruptive 'tornado' hitting some building blocks, to playing energy superheroes, to the fun and games and humour involved in creating plastic totem poles. As I argued, with care, the latter kind of exercise could be seen as a mode of fostering discomfort or dishabituation with our (particularly Minority North) commonplace ways of relating with plastics, whilst drawing attention to other ways of relating with materials, in other contexts.

On the other hand, these forms of silliness exceed speculative fabulation. I have also sought to attend to, develop and exemplify other ways of telling smaller (and sometimes larger) stories about childhoods (after Taylor, 2019). Some of these have involved litany – the actual or metaphorical listing off of objects in order to render them unfamiliar or subject them to thought (after Latour, 2005). Others have involved ontographies or alien phenomenologies (after Bogost, 2012), in which I have attempted to map the pathways of titanium (Figure 8.4) or relate something of what it is *like* to be aluminium or understand the instrumental inter-faces of objects found on an online selling platform with an API. I acknowledge that some readers may, in turn, find (some of) these manoeuvres pointless, uncriti-cal or 'silly'. But there has been a serious point here: inspired by archaeologists of childhood, to bring together uneasy, quirky, perhaps queer alliances of meth-ods and disciplines. Rather than ceding to or celebrating 'the science', these col-laborations offer opportunities to *start-with* a range of ways of knowing (beyond) childhoods that, in turn, enable starting-with plastics and a whole host of other material stuff to tell *other* stories about childhoods than have been possible in the past. Whether 'radical' or not, I have tried to not only promote but exemplify the worth of forms of interdisciplinarity that might enable thinking and doing, after childhood.

Finally, I take the notion of *interface* to have worked in at least two ways in this book. In one sense, it is about the ways in which humans and material stuff do or do not relate with one another – the unit operations, mutually constitutive bathing, digital massification, phasing and interfaces that, inspired by a range of OOO theorists and others, have been developed throughout this book. I do not wish to repeat these here, other than to note once again that they are in part con-stitutive of the ways in which I defined the term 'after' in Chapter 1, and that I retain a sense of scepticism that they can supplant either new materialist or other theoretical approaches in doing childhood studies. Indeed, I have argued in the latter half of the book that arts of (not) noticing should also involve a (re)focusing on children's agency and voice, even if in a quite radically reconfigured light and having started elsewhere. Thus, at the end of the last chapter, I explored how stu-dents' own experiences and mobilities could nuance conceptual languages around *stickiness* and *synthesis* – through attention to modes of habituation, mediation and interaction. Elsewhere, however, I have sought to articulate multiple forms of interfaces – from the conditions that led to *that doll* being found in the apartment in Luz, to encounters with toys and other objects on an online selling platform, to the ways in which aluminium travels into, through and out of children's lives and environments, and the ways in which instrumental interfaces offer different 'senses' of digital traces or elemental traces than our human senses. At times, thinking and doing, *after* childhood, has meant that children have been – pretty much – out of the picture. This is, I think, OK. But as someone still commit-ted to a sense in which there is something about children – socially, legally, biologically, culturally – that means they are, in albeit different ways, generally more vulnerable to trauma than adults, I think it is both possible and politically desirable to hold that 'decentring' (for want of a better word) in tension with or alongside an analysis of *where children are, precisely*. This, then, is why I often emphasised *juxtaposition* (of, for example, totem poles, biosampled data and app-based interviews) as both a form of silliness and a serious response to researching and telling, after childhoods.

On the other hand, and taking a step back, I want to close with a sense in which it is at the point where *trauma and silliness interface* with one another that some of the most productive possibilities emerge. It is in those perhaps local moments (or manifestations) that opportunities for care, love or a post-human, post-child *ethics* and politics emerge (Aitken, 2018). Here, again, for me, OOO, ANT and most brands of new materialism reach their limits, because – as I argued in Chapter 2 – in seeking to extend 'beyond' the human they erase something of the challenging and often traumatic predicaments in which some humans find themselves. Without repeating those critiques, I want to highlight two ways in which moving forwards it might be possible to respond, within and beyond childhood studies. In both cases (and this is very much my reading), there is something about the interface between trauma and silliness that seems to matter, and which underlies many of the analyses in this book.

One way to consider the interface between trauma and silliness is via Berlant's (2011) notion of 'cruel optimism'. Inspired by Marxist, feminist and queer theories, Berlant offers a poignant and unsettling view of Western consumer societies since the Second World War. She argues that many people living in those societies have come to depend on objects (mainly, but not only, the stuff we buy) and ideals (surrounding beauty, family life and democracy, for instance) that ironically actually block our thriving, individually and collectively. We become attached to those objects; we project our hopes and fantasies onto them; yet we are not merely habitually disappointed by these forms of attachment and hope, but they may be damaging to us, to societies and to our planet. Cruel optimism is, thus, like a destructive love affair. Or it is akin to the obesity 'epidemic' (that, incidentally, is considered to be particularly prevalent amongst children), in which the conditions of living and working under capitalism lead us to become reliant on 'treats' to get us through each day – treats that may, ultimately, harm our health. Berlant (2011) asks, therefore, why we become attached (or *stuck*), and why we stay attached, to such forms of cruel optimism. Witnessing what she terms a 'waning of good life genres' – a kind of decline of utopian forms and dispositions – she claims that this is nevertheless a moment to experiment; to work with failure; and for imagining new forms of intimacy.

My reading is that the notion of cruel optimism combines trauma and silliness – quite simply because Berlant's defamiliarisation of the fantasies of consumer capitalism renders them ridiculous, illogical and silly, whilst their *effects* can be pernicious, harmful and traumatic. Thinking of examples in this book, children's attachments to plastics and their circulation and visibility in digital media – which may be both whimsical and toxic, even if not always – would be cases in point. More broadly, however, I wonder if – understood infra-generationally – the positioning of children in many societies – not just in the West – might also be understood as a form of cruel optimism. This would be a corollary of the more familiar sense in which children are understood as 'angels' or 'devils' (Valentine, 1996) – as naturally innocent or evil. However, contemporary childhoods may be understood in the terms of cruel optimism and, especially, an interplay between silliness and trauma. Although not experienced universally, or universally derided, many societies have become attached to ways of imagining childhoods and (in) directly dealing with children that, upon taking a step back, appear at least questionable, if not ridiculous: in the standardisation of testing in education systems; in ever-intensifying digital cultures of visibility; in the dominance of the car in cities; and in the habituated actions and political inactions that propagate rather than challenge climate change. These and many other actions – where children are in or out of focus to greater or lesser extents – can have potentially traumatic effects, even if not always directly or uniformly, and even if not right now: from the rise of mental illness amongst children and young people to the chronic effects of living with air pollution.

Perhaps most controversially, it might be possible to claim that many (Western) societies' relationships with children are cruelly optimistic. Children represent the hopes and dreams of individuals, communities and nations; many of the actions above are taken 'with children's best interests at heart' even though, *at the same time*, we remain attached to programmes like standardised testing or education platforms or AI whilst knowing that they can, as Berlant (2011) puts it, compromise the very conditions for our (and especially children's) living well. All this is to say that I believe that thinking and doing, after childhood, might offer ways to broach such forms of cruel optimism. On the one hand, working through arts of (not) noticing in which children move out of and back into focus, it might be possible to identify and critique forms of silliness/trauma and their effects. Starting with a different focus and starting from a different place (whether geographically, disciplinarily, methodologically) might prompt generative forms of defamiliarisation. In Chapter 8, plastics – and the totem poles – were a key example of this. But working infra-generationally is perhaps key to this approach. It can highlight, for instance, how apparently banal attachments to material stuff have been entangled with socio-economic, environmental, psychological and/or physiological forms of trauma in the past. More simply put, such an approach can help us to learn from childhoods past.

On the other hand, this book has sown the seeds for various ways in which we might imagine childhoods *after* such forms of cruel optimism, and *after* the silliness/trauma of habituations to plastics (etc.). Again, this is not to say that neither childhoods nor notions of childhood will exist in the future. Nor is it to claim that this book offers a blueprint or even a glimpse of 'better' futures for children and childhood. However, in the very same arts of (not) noticing, in speculative fabulations, in litany, in ontography, in alien phenomenology and in playful/ silly encounters *with children and others*, I hope that this book offers a sense of the modes of *experimentation* that Berlant suggests might enable progressive and inclusive forms of thinking and doing, after childhood, to take hold. This may also – although need not necessarily – mean a closer, if careful and critical, attention to what I have termed 'alter-childhoods': to alternative modes of imagining and practising childhoods in unconventionally, radically or experimentally educational, social or economic settings (Kraftl, 2015).

A second way to consider the interface between trauma and silliness is to return to materiality: to interfaces between humans and humans, humans and non-humans, and all kinds of other combinations (what Bryant [2014] terms 'unit operations'). This time, however, and connecting with the more overtly politicised tone of this final chapter, it is to highlight a range of ethical and political questions associated with life itself. Rosi Braidotti (2011) examines the relationship and distinction between bios (as those aspects of life that enter political discourse and social life, and are therefore governed) and zoe (as the vitalistic, prehuman, generative force of life). In this book, zoe corresponds most closely with some of my discussions of energy phenomena, as well as with a sense in which – from speculative realism and OOO in particular – infra-generational life exceeds both individual humans and the human species. Looping back to the feminist, critical race and queer theoretical strands that have been woven throughout this book, Braidotti also argues that zoe marks the constitutive outside of the predominant white, Western, male mode of doing and structuring life. She argues, for instance, that the rise of genetic engineering witnesses an epoch in which we have become 'infrahuman' (Braidotti, 2011: 327) – wherein bios and zoe have become muddled, and zoe has become splintered. Noting the resonance with my concept of infra-generations, this muddling (and meddling) extends to many other domains, captured (if not satisfactorily) by the notion of the Anthropocene. It is, most importantly, as Bennett (2010) also argues, a recognition of us all being in this predicament together – bios and zoe, human and non-human. Moreover, as Morton (2013) writes of hyperobjects like climate change, this predicament is precisely one of *silliness* (dominant humans' lameness and hypocrisy) and a potential ecological *trauma* yet-to-come that is only beginning to phase into human modes of perception and measurement.

Braidotti's (2011: 327) response is to engage in ecophilosophical ethics of sustainability that open up "alternative ecologies of belonging" and which require forms of entanglement beyond those centred upon humans. Unlike Aitken (2018), however, whose wonderful analysis of the post-child signals fundamental implications for how we might view children's rights in an ecophilosophical manner, I find Braidotti's writings on *death* an appropriate and hopefully provocative place to end this book. Inspired by Deleuze, Braidotti continually seeks to find ways in which humans might become other. The *ultimate* 'other' of human life is death. Death is the anterior of life (the after-life): and life is, whether from quotidian, scientific or a range of spiritual viewpoints, founded upon death. The earthly presence of the bones of children, and fossils, and oil, reminds us of a time anterior to (modern) human existence and perception. Death is also, as Braidotti (2011) argues, an ever-present condition of life and human being: we are always aware of our own death. In both senses (past *and* future), death reminds us of human finitude so that – as per Meillassoux (2010) – the true meaning of the term 'after finitude' denotes *both* historicity and futurity, not just a linear (con)sequence. As per my understanding of the term 'after' at the beginning of the book, and my deliberate use of historical and archaeological materials in Chapter 6, the term 'after' offers a more muddled, complex, sense of temporality than even the prefix 'post-' allows.

Braidotti (2011: 328) poses a significant question with which I would like to end this book: how to "rethink ethical and political life beyond survival and mortality"? She is less interested in what might happen once an individual life ends, because this implies an overemphasis on death as a condition of the life of the individuated subject as understood in egocentric Western cultures. Rather than personal death, then, Braidotti seeks to think through *im*personal death: a context of life without either 'me' or *any* human. In many ways, this represents the ultimate 'decentring' of humans. It is situated beyond, even, the 'entanglements' of the new materialists but – because death is in this chronology always *both* behind and ahead of humans – is a way of imagining life *after* humans. As Haraway (2011) puts it otherwise, in explaining speculative fabulation, this is a multi-generational, more-than-human (or infra-generational) ethic of care that extends kinship to the distant past *and* distant futures. For Braidotti, this is a mode of thinking with, not against death. The latter sets life and death in a dualistic, teleological logic and has afforded the conditions for powerful white, Western men to set the conditions of life (and death) for the majority, including the many children clinging to life as, in Katz's (2018) terms, the 'waste of the world'. The former – a reclaiming of death and a subversion of death's socio-ecological status – could be the ultimate threat to the silliness/trauma writ by the latter.

Just as Berlant (2011) sees experimentation and working with failure as a possible method for addressing cruel optimism, Braidotti (2011) asks whether a range of (especially contemporary) human conditions that are marked by trauma *and* silliness might offer modes for thinking-with death. She argues that conditions like alcoholism and other forms of substance dependency, anorexia and a range of behavioural conditions have, rather than exceptions, become "markers of a standard condition, namely, the human subjects' enfleshed exposure to the irrepressible and at time hurtful vitality of life" (Braidotti, 2011: 342). Once again, these are all conditions that affect children in various ways (and the evidence shows, generally, that their prevalence is increasing). And once again the term 'silliness' is not meant flippantly or dismissively but to chart a range of effects and affects associated with these conditions: to be sure, 'silliness' may be an outcome of drunkenness, just as in very different ways it might be a feature of teenage boys' emotional, social or behavioural differences and the impulse to call a tree a 'gas-guzzling mother-fucker'. But these addictions, illnesses and behaviours are also associated (particularly amongst those who do not experience them) with being illogical, ridiculous and pointless, just as much as they are traumatic, hurtful, damaging and, in some cases, potentially lethal. Nevertheless, Braidotti (2011: 344) also sees in these conditions – perhaps somewhat controversially – opportunities for other modes of interaction, where those who have "already cracked up a bit" can open up cracks in habituated modes of being that might lead to ethical transformations or engender sites where values can be transposed.

As elsewhere in the book, I am not seeking to endorse or romanticise (especially) forms of behaviour that are often marked and then simply dismissed as the 'silliness' of teenage boys (or those other, Others listed above) who do not conform. However, neither do I want to condemn them or silence them. There is another path: working- and thinking-with such forms of silliness/trauma, both in asking deeper questions about the socio-ecological contexts that have allowed them and in listening to and acting upon the demands that are expressed by excluded students in Balsall Heath as much as the progenitors of the #climatestrikes.

The implications for thinking and doing, after childhood, are potentially profound, and I close by listing a few of these by way of a challenge and provocation for future work in and beyond childhood studies. First, taken to their extreme, imagining contexts of life *after* the human inevitably bears implications not only for thinking-with death but how we think about fertility. Indeed, one response to climate change (but also to other social questions) has been to imagine futures in which human fertility is reduced – in which some humans choose not to have children (known as antinatalism). If the aim would be to generate the conditions for life after humanity, then that would, quite literally, imply a world *after childhoods*. Second, although less extreme in terms of intent, queer theorists have, amongst others, pointed to a sense in which humanity may well be on an inadvertent path to self-destruction that is an effect of its cruel optimism about fossil fuels, plastics, the motorcar and the other trappings of consumer cultures. As Davis (2015), amongst others, points out, plastics are one of several materials that are (in my terms) *synthesising* with human bodies and *sticking*-with the earth in ways that are altering the conditions of fertility and of being a child: plasticisers, like other toxins, reducing sperm counts in men (further queer(y)ing assumed sexual binaries) whilst altering the 'ordinary' development of children's and adolescents' brains. Given the temporal undulations (Morton, 2013) associated with plastics, cements and other human-made hyperobjects, even if not on a path to the ultimate death of the human, the human species (and the earth it inhabits centuries from now) may look utterly different as the effects of our contemporary chemical kinships accumulate and wreak their effects on the genetic codes, behaviours and physiologies of future generations. Thus – and for some commentators the seeds are already with us, in the present – we may if not deliberately (as per the antinatalists) but accidentally be heading towards an earth *after* the human as understood now and, as a result, necessarily *after* childhoods as they are understood now, and may need to learn how to live with these 'queer kin' (Haraway, 2011).

I raise both of the above scenarios *not* to endorse them in any sense but to provoke and challenge. These modes of thinking about fertility, human life and therefore childhood are just two of many that challenge especially contemporary Western understandings of childhood and which certainly require critical, careful and thoughtful interrogation by childhood studies scholars. Alongside the many other ways in which I have (not) noticed children and childhoods in this book, both offer, in Braidotti's (2011: 333) terms, "relentlessly generative" ways of engaging with life and death (and with silliness/trauma, depending on how one feels about them) in ways that are conceived as human, non-human, "inhuman and as post-human". Whether one agrees with them or not, childhoods would weave in and out of focus and child*ren* would be conceived not as the singular, bounded subject to be moulded in the likeness of our cruel optimisms but held in relation to the ways in which more-than-human ethics of responsibility might grapple, infra-generationally, with our past and present silliness/trauma, and imagine other ways of being. Those other ways of being may not be quite as extreme as the two provocations listed above, although they both (and especially the latter) require serious attention, not least for their implications for children, both now and in the future. Nonetheless, thinking-with these and other scenarios also requires modes of inter-disciplinary critique, experimentation, narrative and speculation that I have tried to theorise and exemplify in this book. In other words, they require thinking and doing, *after childhood*.

References

Agard-Jones, V. (2012) What the sands remember. *GLQ: A Journal of Lesbian and Gay Studies*, 18: 325–346.

Agard-Jones, V. (2013) Bodies in the system. *Small Axe: A Caribbean Journal of Criticism*, 17: 182–192.

AGHS (Acock's Green History Society) (2019) *Brick and tile making in Yardley*. Available online at: https://aghs.jimdo.com/brick-and-tile-making/, last accessed 17th October 2019.

Aguirre-Bielschowsky, I., Lawson, R., Stephenson, J. and Todd, S. (2017) Energy literacy and agency of New Zealand children. *Environmental Education Research*, 23: 832–854.

Aitken, S.C. (2010) Bold disciplinarianism, experimentation and failing spectacularly. *Children's Geographies*, 8: 219–220.

Aitken, S.C. (2018) *Young people, rights and place: Erasure, neoliberal politics and post-child ethics*. London: Routledge.

Aitken, S.C. and Herman, T. (1997) Gender, power and crib geography: Transitional spaces and potential places. *Gender, Place and Culture: A Journal of Feminist Geography*, 4: 63–88.

Alanen, L. and Mayall, B. eds. (2001) *Conceptualizing child-adult relations*. London: Psychology Press.

Allouche, J., Middleton, C. and Gyawali, D. (2015) Technical veil, hidden politics: Interrogating the power linkages behind the nexus. *Water Alternatives*, 8: 61–626.

Álvarez, H.H. (2017) Childhood and material culture at Hacienda San Pedro Cholul during Yucatán's gilded age. *Childhood in the Past*, 10: 122–141.

Anderson, B. (2010) Preemption, precaution, preparedness: Anticipatory action and future geographies. *Progress in Human Geography*, 34: 777–798.

Anderson, B. and McFarlane, C. (2011) Assemblage and geography. *Area*, 43: 124–127.

Andres, L. (2012) Differential spaces, power hierarchy and collaborative planning: A critique of the role of temporary uses in shaping and making places. *Urban Studies*, 50: 759–775.

Änggård, E. (2015) Digital cameras: Agents in research with children. *Children's Geographies*, 13: 1–13.

Ansell, N. (2009) Childhood and the politics of scale: Descaling children's geographies? *Progress in Human Geography*, 33: 190–209.

Ansell, N. (2016) *Children, youth and development*. London: Routledge.

Ansell, N., Hajdu, F., van Blerk, L. and Robson, E. (2019) Fears for the future: The incommensurability of securitisation and in/securities among southern African youth. *Social & Cultural Geography*, 20: 507–533.

Ash, J., Anderson, B., Gordon, R. and Langley, P. (2018) Unit, vibration, tone: A post-phenomenological method for researching digital interfaces. *Cultural Geographies*, 25: 165–181.

Ash, J. and Simpson, P. (2016) Geography and post-phenomenology. *Progress in Human Geography*, 40: 48–66.

Balagopalan, S. (2019) 'Afterschool and during vacations': On labor and schooling in the postcolony. *Children's Geographies*, 17: 231–245.

Balagopalan, S., Coe, C. and Green, K.M., eds. (2019) *Diverse unfreedoms: The afterlives and transformations of post-transatlantic bondages*. London: Routledge.

Barad, K. (2007) *Meeting the universe halfway: Quantum physics and the entanglement of matter and meaning*. Durham: Duke University Press.

Bauer, I. (2015) Approaching geographies of education differANTly. *Children's Geographies*, 13: 620–627.

BBC (2016) Syrian children burn tyres to create no-fly zones. Available online at: www.bbc.co.uk/news/world-36944470, last accessed 6th August 2019.

BBC (2018a) Anti-plastic focus 'dangerous distraction' from climate change. Available online at: www.bbc.co.uk/news/uk-45942814, last accessed 6th November 2019.

BBC (2018b) Amazon and eBay pull CloudPets smart toys from sale. Available online at: www.bbc.co.uk/news/technology-44382135, last accessed 15th August 2019.

BBC (2019a) Amazon sued over Alexa child recordings in US. Available online at: www.bbc.co.uk/news/technology-48623914, last accessed 15th August 2019.

BBC (2019b) Burger King ditches free toys and will 'melt' old ones. Available online at: www.bbc.co.uk/news/business-49738889, last accessed 1st October 2019.

Beck, A., Chitalia, S. and Rai, V. (2019) Not so gameful: A critical review of gamification in mobile energy applications. *Energy Research and Social Science*, 51: 32–39.

Belontz, S.L., Corcoran, P.L., Davis, H., Hill, K.A., Jazvac, K., Robertson, K. and Wood, K. (2019) Embracing an interdisciplinary approach to plastics pollution awareness and action. *Ambio*, 48: 855–866.

Bengston, J.D. and O'Gorman, J.A. (2016) Children, migration and mortuary representation in the late prehistoric Central Illinois River Valley. *Childhood in the Past*, 9: 19–43.

Bennett, J. (2001) *The enchantment of modern life: Attachments, crossings, and ethics*. Boston: Princeton University Press.

Bennett, J. (2010) *Vibrant matter: A political ecology of things*. Durham, NC: Duke University Press.

Bergmann, M., Mützel, S., Primpke, S., Tekman, M.B., Trachsel, J. and Gerdts, G. (2019) White and wonderful? Microplastics prevail in snow from the Alps to the Arctic. *Science Advances*, 5: unpaginated.

Berlant, L.G. (2011) *Cruel optimism*. Durham, NC: Duke University Press.

Biran, A., Abbot, J. and Mace, R. (2004) Families and firewood: A comparative analysis of the costs and benefits of children in firewood collection and use in two rural communities in sub-Saharan Africa. *Human Ecology*, 32: 1–25.

Blaise, M., Hamm, C. and Iorio, J.M. (2017) Modest witness(ing) and lively stories: Paying attention to matters of concern in early childhood. *Pedagogy, Culture & Society*, 25: 31–42.

Blum-Ross, A., Donoso, V., Dinh, T., Mascheroni, G., O'Neill, B., Riesmeyer, C. and Stoilova, M. (2018) *Looking forward: Technological and social change in the lives of European children and young people*. Brussels: ICT Coalition.

Bogost, I. (2006) *Unit operations: An approach to videogame criticism*. Cambridge, MA: MIT Press.

Bogost, I. (2012) *Alien phenomenology, or, what it's like to be a thing.* Minneapolis: University of Minnesota Press.

Bogost, I. (2015) *How to talk about videogames.* Minneapolis: University of Minnesota Press.

Boing Boing (2017) Boaters stumble on massive Caribbean trash vortex. Available online at: https://boingboing.net/2017/11/20/boaters-stumble-on-massive-car.html, last accessed 7th November 2019.

Bond, E. (2014) *Childhood, mobile technologies and everyday experiences: Changing technologies= changing childhoods?* Berlin: Springer.

Bourdier, P., Piponnier, E., Blazevich, A., Maciejewski, H., Duche, P. and Ratel, S. (2018) Metabolic and fatigue profiles are comparable between prepubertal children and well-trained adult endurance athletes. *Frontiers in Physiology*, 9: unpaginated.

Boyd, D. (2014) *It's complicated: The social lives of networked teens.* New Haven: Yale University Press.

Bradbury, A. (2019) Datafied at four: The role of data in the 'schoolification' of early childhood education in England. *Learning, Media and Technology*, 44: 7–21.

Braidotti, R. (2011) *Nomadic theory: The portable Rosi Braidotti.* New York: Columbia University Press.

Bravo-Marquez, F., Mendoza, M. and Poblete, B. (2014) Meta-level sentiment models for big social data analysis. *Knowledge-Based Systems*, 69: 86–99.

Brickell, K., Cristofoletti, T., Chann, S., Natarajan, N. and Parsons, L. (2019) Exhibition: Blood bricks: Untold stories of modern slavery and climate change from Cambodia. *Environment and Planning D: Society and Space Blog.* Available online at: https://societyandspace.org/2019/03/27/blood-bricks-untold-stories-of-modern-slavery-and-climate-change-from-cambodia/, last accessed 17th October 2019.

Broto, V.C., Allen, A. and Rapoport, E. (2012) Interdisciplinary perspectives on urban metabolism. *Journal of Industrial Ecology*, 16: 851–861.

Brown, G. and Kraftl, P. (2019, online early) Theorising cohortness: (Mis)fitting into student geographies. *Transactions of the Institute of British Geographers*, 44: 616–632.

Brown, G. and Pickerill, J. (2009) Space for emotion in the spaces of activism. *Emotion, Space and Society*, 2: 24–35.

Bryant, L.R. (2014) *Onto-cartography.* Edinburgh: Edinburgh University Press.

Büntgen, U., Tegel, W., Nicolussi, K., McCormick, M., Frank, D., Trouet, V., Kaplan, J.O., Herzig, F., Heussner, K.U., Wanner, H. and Luterbacher, J. (2011) 2500 years of European climate variability and human susceptibility. *Science*, 331: 578–582.

Burnett, J. (2016) *Generations: The time machine in theory and practice.* London: Routledge.

Cairns, R. and Krzywoszynska, A. (2016) Anatomy of a buzzword: The emergence of 'the water-energy-food nexus' in UK natural resource debates. *Environmental Science & Policy*, 64: 164–170.

Callon, M. (1986) The sociology of an actor-network: The case of the electric vehicle. In *Mapping the dynamics of science and technology* (pp. 19–34). London: Palgrave Macmillan.

Cessford, C. (2017) Throwing away everything but the kitchen sink? Large assemblages, depositional practice and post-medieval households in Cambridge. *Post-Medieval Archaeology*, 51: 164–193.

Chang, K.H., Hsu, P.Y., Lin, C.J., Lin, C.L., Juo, S.H.H. and Liang, C.L. (2019) Traffic-related air pollutants increase the risk for age-related macular degeneration. *Journal of Investigative Medicine*, 67: 1076–1081.

Chen, M.Y. (2011) Toxic animacies, inanimate affections. *GLQ: A Journal of Lesbian and Gay Studies*, 17: 265–286.

Choudhury, S. and McKinney, K.A. (2013) Digital media, the developing brain and the interpretive plasticity of neuroplasticity. *Transcultural Psychiatry*, 50: 192–215.

Clark, J., Gurung, P., Chapagain, P.S., Regmi, S., Bhusal, J.K., Karpouzoglou, T., Mao, F. and Dewulf, A. (2017) Water as 'time-substance': The hydrosocialities of climate change in Nepal. *Annals of the American Association of Geographers*, 107: 1351–1369.

Collins, R.C. (2014) *Excessive . . . but not wasteful? Exploring young people's material consumption through the lens of divestment* (Doctoral dissertation, UCL, University College London, London).

Cook, I. (2004) Follow the thing: Papaya. *Antipode*, 36: 642–664.

Corcoran, P.L., Moore, C.J. and Jazvac, K. (2014) An anthropogenic marker horizon in the future rock record. *GSA Today*, 24: 4–8.

Crawford, S. (2009) The archaeology of play things: Theorising a toy stage in the 'biography' of objects. *Childhood in the Past*, 2: 55–70.

Crewe, V.A. and Hadley, D.M. (2013) 'Uncle Tom was there, in crockery': Material culture and a Victorian working-class childhood. *Childhood in the Past*, 6: 89–105.

Crinall, S. and Somerville, M. (2019) Informal environmental learning: The sustaining nature of daily child/water/dirt relations. *Environmental Education Research*, 2019: 1–12.

Cukurova, M., Kent, C. and Luckin, R. (2019, online early) Artificial Intelligence and multimodal data in the service of human decision-making: A case study in debate tutoring. *British Journal of Educational Technology*.

Davies, T. (2019, online early) Slow violence and toxic geographies: 'Out of sight' to whom? *Environment and Planning C: Politics and Space*.

Davis, H. (2015) Life & death in the Anthropocene: A short history of plastic. In *Art in the Anthropocene: Encounters among aesthetics, politics, environments and epistemologies* (pp. 347–358). London: Open Humanities Press.

Day, C. (2016) Education and employment transitions: The experiences of young people with caring responsibilities in Zambia. In *Labouring and learning* (pp. 1–26). Berlin: Springer.

de Campos Tebet, G. and Abramowicz, A. (2018) Estudos de bebês: linhas e perspectivas de um campo em construção. *ETD-Educação Temática Digital*, 20: 924–946.

De Hoop, E. (2017) Multiple environments: South Indian children's environmental subjectivities in formation. *Children's Geographies*, 15: 570–582.

DeLanda, M. (2019) *A new philosophy of society: Assemblage theory and social complexity*. London: Bloomsbury Publishing.

Deleuze, G. and Guattari, F. (1988) *A thousand plateaus: Capitalism and schizophrenia*. London: Bloomsbury Publishing.

De Loughry, T. (2019) Polymeric chains and petrolic imaginaries: World literature, plastic, and negative value. *Green Letters*, 23: 179–193.

den Besten, O., Horton, J., Adey, P. and Kraftl, P. (2011) Claiming events of school (re) design: Materialising the promise of building schools for the future. *Social & Cultural Geography*, 12: 9–26.

DeWaters, J. and Powers, S. (2011) Improving energy literacy among middle school youth with project-based learning pedagogies. *Frontiers in Education Conference Proceedings*, unpaginated.

Dixon, S.J., Viles, H.A. and Garrett, B.L. (2018) Ozymandias in the Anthropocene: The city as an emerging landform. *Area*, 50: 117–125.

Dozier, C.A. (2016) Finding children without toys: The archaeology of children at Shabbona Grove, Illinois. *Childhood in the Past*, 9: 58–74.

Dris, R., Gasperi, J., Saad, M., Mirande, C. and Tassin, B. (2016) Synthetic fibers in atmospheric fallout: A source of microplastics in the environment? *Marine Pollution Bulletin*, 104: 290–293.

Ember, C.R. and Cunnar, C.M. (2015) Children's play and work: The relevance of cross-cultural ethnographic research for archaeologists. *Childhood in the Past*, 8: 87–103.

Endo, A., Burnett, K., Orencio, P., Kumazawa, T., Wada, C., Ishii, A., Tsurita, I. and Taniguchi, M. (2015) Methods of the water-energy-food nexus. *Water*, 7: 5806–5830.

EPA (2019) *Human health and climate change research*. Available online at: https://19january2017snapshot.epa.gov/climate-impacts/climate-impacts-human-health_.html, last accessed 15th October 2019.

Ergler, C.R., Kearns, R., Witten, K. and Porter, G. (2016) Digital methodologies and practices in children's geographies. *Children's Geographies*, 14: 129–140.

Esson, J. (2015) Better off at home? Rethinking responses to trafficked West African footballers in Europe. *Journal of Ethnic and Migration Studies*, 41: 512–530.

Evans, B. (2010) Anticipating fatness: Childhood, affect and the pre-emptive 'war on obesity'. *Transactions of the Institute of British Geographers*, 35: 21–38.

Evans, R. and Thomas, F. (2009) Emotional interactions and an ethics of care: Caring relations in families affected by HIV and AIDS. *Emotion, Space and Society*, 2: 111–119.

Feinstein, N.W., Jacobi, P.R. and Lotz-Sisitka, H. (2013) When does a nation-level analysis make sense? ESD and educational governance in Brazil, South Africa, and the USA. *Environmental Education Research*, 19: 218–230.

Fell, M. and Fong Chiu, L. (2014) Children, parents and home energy use: Exploring motivations and limits to energy demand reduction. *Energy Policy*, 65: 351–358.

Finn, M. (2016) Atmospheres of progress in a data-based school. *Cultural Geographies*, 23: 29–49.

Franceschi-Bicchierai, L. (2017) Internet of things teddy bear leaked 2 million parent and kids message recordings. *Motherboard: Tech by Vice*. Available online at: www.vice.com/en_us/article/pgwean/internet-of-things-teddy-bear-leaked-2-million-parent-and-kids-message-recordings, last accessed 15th August 2019.

Franklin, S., Grey, M.J., Heneghan, N., Bowen, L. and Li, F.X. (2015) Barefoot vs common footwear: A systematic review of the kinematic, kinetic and muscle activity differences during walking. *Gait & Posture*, 42: 230–239.

Fulminante, F. (2015) Infant feeding practices in Europe and the Mediterranean from prehistory to the middle ages: A comparison between the historical sources and bioarchaeology. *Childhood in the Past*, 8: 24–47.

Gagen, E.A. (2015) Governing emotions: Citizenship, neuroscience and the education of youth. *Transactions of the Institute of British Geographers*, 40: 140–152.

Gallagher, M. (2019, online early) Childhood and the geology of media. *Discourse: Studies in the Cultural Politics of Education*.

Georgiadis, M. (2011) Child burials in Mesolithic and Neolithic southern Greece: A synthesis. *Childhood in the Past*, 4: 31–45.

Ghosh, R. (2019) Plastic literature. *University of Toronto Quarterly*, 88: 277–291.

Gibson-Graham, J.K. (2006) *A postcapitalist politics*. Minneapolis: University of Minnesota Press.

Goldthau, A. (2014) Rethinking the governance of energy infrastructure: Scale, decentralization and polycentrism. *Energy Research & Social Science*, 1: 134–140.

Goodyear, V. and Armour, K. (2018) *Young people, social media and health*. London: Routledge.

Gowland, R. and Redfern, R. (2010) Childhood health in the Roman World: Perspectives from the centre and margin of the Empire. *Childhood in the Past*, 3: 15–42.

Gratton, P. (2014) *Speculative realism: Problems and prospects*. London: Bloomsbury Publishing.

Gregory, M. (2009) Environmental implications of plastic debris in marine settings: Entanglement, ingestion, smothering, hangers-on, hitch-hiking and alien invasions. *Philosophical Transactions*, 364: 2013–2025.

Gregson, N., Metcalfe, A. and Crewe, L. (2007) Moving things along: The conduits and practices of divestment in consumption. *Transactions of the Institute of British Geographers*, 32: 187–200.

Grosz, E. (2011) *Becoming undone: Darwinian reflections on life, politics, and art*. Durham: Duke University Press.

Guardian Newspaper (2019) Growing up in air-polluted areas linked to mental health issues. Available online at: www.theguardian.com/society/2019/aug/20/growing-up-in-air-polluted-areas-linked-to-mental-health-issues, last accessed 18th October 2019.

Gulson, K.N. and Webb, P.T. (2017) Mapping an emergent field of 'computational education policy': Policy rationalities, prediction and data in the age of Artificial Intelligence. *Research in Education*, 98: 14–26.

Gulson, K.N. and Sellar, S. (2019). Emerging data infrastructures and the new topologies of education policy. *Environment and Planning D: Society and Space*, 37(2): 350–366.

Gustafson, S., Heynen, N., Rice, J.L., Gragson, T., Shepherd, J.M. and Strother, C. (2014) Megapolitan political ecology and urban metabolism in Southern Appalachia. *The Professional Geographer*, 66: 664–675.

Hackett, A. and Rautio, P. (2019) Answering the world: Young children's running and rolling as more-than-human multimodal meaning making. *International Journal of Qualitative Studies in Education*, 32: 1019–1031.

Hadfield-Hill, S. and Zara, C. (2019, online early) Children and young people as geological agents? Time, scale and multispecies vulnerabilities in the new epoch. *Discourse: Studies in the Cultural Politics of Education*.

Halberstam, J. (2005) *In a queer time and place: Transgender bodies, subcultural lives*. New York: NYU press.

Hansen, S.R., Hansen, M.W. and Kristensen, N.H. (2017) Striated agency and smooth regulation: Kindergarten mealtime as an ambiguous space for the construction of child and adult relations. *Children's Geographies*, 15: 237–248.

Haraway, D. (1997) *Modest_Witness@Second_Millennium. FemaleMan_Meets_OncoMouse: Feminism and technoscience*. London: Routledge.

Haraway, D. (2011) Speculative fabulations for technoculture's generations: Taking care of unexpected country. *Australian Humanities Review*, 50: 1–18.

Haraway, D. (2016) *Staying with the trouble: Making kin in the Chthulucene*. Durham: Duke University Press.

Harman, G. (2010a) *Towards speculative realism: Essays and lectures*. Alresford: John Hunt Publishing.

Harman, G. (2010b) *Prince of networks: Bruno Latour and metaphysics*. Melbourne: re.press.

Harman, G. (2011a) *Tool-being: Heidegger and the metaphysics of objects*. Chicago: Open Court.

Harman, G. (2011b) *Guerrilla metaphysics: Phenomenology and the carpentry of things.* Chicago: Open Court.

Harper, E. (2018) Toys and the portable antiquities scheme: A source for exploring later medieval childhood in England and Wales. *Childhood in the Past*, 11: 85–99.

Harper, K. (2017) *The fate of Rome: Climate, disease, and the end of an empire.* Princeton, NJ: Princeton University Press.

Haunstrup Christensen, T. and Rommes, E. (2019) Don't blame the youth: The social-institutional and material embeddedness of young people's energy-intensive use of information and communication technology. *Energy Research & Social Science*, 49: 82–90.

Himelboim, I., Smith, M.A., Rainie, L., Shneiderman, B. and Espina, C. (2017) Classifying Twitter topic-networks using social network analysis. *Social Media+ Society*, 3: 1–13.

Hodgins, D. (2019) *Feminist research for 21st-century childhoods: Common worlds methods.* London: Bloomsbury Publishing.

Hoff, H. (2011) *Understanding the nexus: Background paper for the Bonn 2011 conference.* Bonn: UNEP.

Holloway, J. (2010) *Crack capitalism.* London: Pluto Press.

Holloway, S.L. (1998) Local childcare cultures: Moral geographies of mothering and the social organisation of pre-school education. *Gender, Place and Culture: A Journal of Feminist Geography*, 5: 29–53.

Holloway, S.L., Holt, L. and Mills, S. (2019) Questions of agency: Capacity, subjectivity, spatiality and temporality. *Progress in Human Geography*, 43: 458–477.

Holloway, S.L. and Valentine, G. (2000) Spatiality and the new social studies of childhood. *Sociology*, 8, 34: 763–783.

Holloway, S.L. and Valentine, G. (2001) 'It's only as stupid as you are': Children's and adults' negotiation of ICT competence at home and at school. *Social & Cultural Geography*, 2: 25–42.

Holt, L. (2013) Exploring the emergence of the subject in power: Infant geographies. *Environment and Planning D: Society and Space*, 31: 645–663.

Holt, L. (2017) Food, feeding and the material everyday geographies of infants: Possibilities and potentials. *Social & Cultural Geography*, 18: 487–504.

hooks, B. (2003) *Teaching community: A pedagogy of hope.* London: Psychology Press.

Hoolachan, J. and McKee, K. (2019) Inter-generational housing inequalities: 'Baby boomers' versus the 'millennials'. *Urban Studies*, 56: 210–225.

Hopkins, P. and Pain, R. (2007) Geographies of age: Thinking relationally. *Area*, 39: 287–294.

Horn, M., Atrash Leong, Z., Greenberg, M. and Stevens, R. (2015) Kids and thermostats: Understanding children's involvement with household energy systems. *International Journal of Child-Computer Interaction*, 3: 14–22.

Horton, J. (2010) 'The best thing ever': How children's popular culture matters. *Social & Cultural Geography*, 11: 377–398.

Horton, J. (2012) 'Got my shoes, got my Pokémon': Everyday geographies of children's popular culture. *Geoforum*, 43: 4–13.

Horton, J., Hadfield-Hill, S. and Kraftl, P. (2015) Children living with 'sustainable' urban architectures. *Environment and Planning A*, 47: 903–921.

Horton, J. and Kraftl, P. (2005) For more-than-usefulness: Six overlapping points about children's geographies. *Children's Geographies*, 3: 131–143.

Horton, J. and Kraftl, P. (2006a) What else? Some more ways of thinking and doing 'children's geographies'. *Children's Geographies*, 4: 69–95.

Horton, J. and Kraftl, P. (2006b) Not just growing up, but going on: Materials, spacings, bodies, situations. *Children's Geographies*, 4: 259–276.

Horton, J. and Kraftl, P. (2008) Reflections on geographies of age: A response to Hopkins and Pain. *Area*, 40: 284–288.

Horton, J. and Kraftl, P. (2018) Rats, assorted shit and 'racist groundwater': Towards extra-sectional understandings of childhoods and social-material processes. *Environment and Planning D: Society and Space*, 36: 926–948.

Huang, M.N. (2017) Ecologies of entanglement in the great pacific garbage patch. *Journal of Asian American Studies*, 20: 95–117.

Huijsmans, R. (2016) *Generationing development: A relational approach to children, youth and development*. Berlin: Springer.

Hultman, K. and Lenz Taguchi, H. (2010) Challenging anthropocentric analysis of visual data: A relational materialist methodological approach to educational research. *International Journal of Qualitative Studies in Education*, 23: 525–542.

Intergenerational Commission (2018) *A new generational contract*. London: Resolution Foundation.

Ito, M., Baumer, S., Bittanti, M., boyd, d. and Cody, R. (2010) *Hanging out, messing around, and geeking out: Kids living and learning with new media*. Cambridge, MA: MIT Press.

Jackson, Z.I. (2015) Outer worlds: The persistence of race in movement 'beyond the human'. *GLQ: A Journal of Lesbian and Gay Studies*, 21: 215–218.

Jambeck, J.R., Geyer, R., Wilcox, C., Siegler, T.R., Perryman, M., Andrady, A., Narayan, R. and Law, K.L. (2015) Plastic waste inputs from land into the ocean. *Science Magazine*, 347: 768–771.

Jarke, J. and Breiter, A. (2019) Editorial: The datafication of education. *Learning, Media and Technology*, 44: 1–6.

Jay, M. (2009) Breastfeeding and weaning behaviour in archaeological populations: Evidence from the isotopic analysis of skeletal materials. *Childhood in the Past*, 2: 163–178.

Jeffrey, C. (2013) Geographies of children and youth III: Alchemists of the revolution? *Progress in Human Geography*, 37: 145–152.

Johnson, S. (2019) Better representation in artificial intelligence starts early. *EdSurge Website*. Available online at: www.edsurge.com/news/2019-05-21-better-representation-in-artificial-intelligence-starts-early, last accessed 15th August 2019.

Jonah, O.T. and Abebe, T. (2019) Tensions and controversies regarding child labor in small-scale gold mining in Ghana. *African Geographical Review*, 38: 361–373.

Jones, O. (2008) 'True geography [] quickly forgotten, giving away to an adult-imagined universe': Approaching the otherness of childhood. *Children's Geographies*, 6: 195–212.

Kallio, K.P. and Häkli, J. (2013) Children and young people's politics in everyday life. *Space and Polity*, 17: 1–16.

Katz, C. (1996) Towards minor theory. *Environment and Planning D: Society and Space*, 14: 487–499.

Katz, C. (2004) *Growing up global: Economic restructuring and children's everyday lives*. Minneapolis: University of Minnesota Press.

Katz, C. (2018) The Angel of Geography: Superman, Tiger Mother, aspiration management, and the child as waste. *Progress in Human Geography*, 42: 723–740.

Kelly, P. and Kamp, A. (2015) *A critical youth studies for the 21st century*. Leiden: Brill.

Khan, A., Plana-Ripoll, O., Antonsen, S., Brandt, J., Geels, C., Landecker, H., Sullivan, P.F., Pedersen, C.B. and Rzhetsky, A. (2019) Environmental pollution is associated with increased risk of psychiatric disorders in the US and Denmark. *PLoS Biology*, 17: unpaginated.

Klein, J. (2019) The moment we started using plastic is preserved forever in the planet's fossil record. Available online at: www.fastcompany.com/90400111/the-moment-we-started-using-plastic-is-preserved-forever-in-the-planets-fossil-record, last accessed 7th November 2019.

Koelmans, A.A., Nor, N.H.M., Hermsen, E., Kooi, M., Mintenig, S.M. and De France, J. (2019) Microplastics in freshwaters and drinking water: Critical review and assessment of data quality. *Water Research*, 155: 410–422.

Konstantoni, K. and Emejulu, A. (2017) When intersectionality met childhood studies: The dilemmas of a travelling concept. *Children's Geographies*, 15: 6–22.

Kraftl, P. (2007) Utopia, performativity, and the unhomely. *Environment and Planning D: Society and Space*, 25: 120–143.

Kraftl, P. (2008) Young people, hope, and childhood-hope. *Space and Culture*, 11: 81–92.

Kraftl, P. (2010) Architectural movements, utopian moments: (In)coherent renderings of the Hundertwasser-Haus, Vienna. *Geografiska Annaler: Series B, Human Geography*, 92: 327–345.

Kraftl, P. (2013a) Beyond 'voice', beyond 'agency', beyond 'politics'? Hybrid childhoods and some critical reflections on children's emotional geographies. *Emotion, Space and Society*, 9: 13–23.

Kraftl, P. (2013b) *Geographies of alternative education: Diverse learning spaces for children and young people*. Bristol: Policy Press.

Kraftl, P. (2014) Liveability and urban architectures: Mol(ecul)ar biopower and the 'becoming lively' of sustainable communities. *Environment and Planning D: Society and Space*, 32: 274–292.

Kraftl, P. (2015) Alter-childhoods: Biopolitics and childhoods in alternative education spaces. *Annals of the Association of American Geographers*, 105: 219–237.

Kraftl, P. (2018) A double-bind? Taking new materialisms elsewhere in studies of education and childhood. *Research in Education*, 101: 30–38.

Kraftl, P., Balestieri, J.A.P., Campos, A.E.M., Coles, B., Hadfield-Hill, S., Horton, J., Soares, P.V., Vilanova, M.R.N., Walker, C. and Zara, C. (2019) (Re)thinking (re)connection: Young people, 'natures' and the water-energy-food nexus in São Paulo State, Brazil. *Transactions of the Institute of British Geographers*, 44: 299–314.

Kraftl, P., Hadfield-Hill, S. and Lynch, I. (forthcoming) So you're *literally* taking the p**s?! Plastic ethics: Critically analysing and accounting for ethics and risk in interdisciplinary research on children and plastics. *Children's Geographies*.

Kullman, K. (2012) Experiments with moving children and digital cameras. *Children's Geographies*, 10: 1–16.

Kullman, K. and Palludan, C. (2011) Rhythmanalytical sketches: Agencies, school journeys, temporalities. *Children's Geographies*, 9: 347–359.

Kumar, P.C., Vitak, J., Chetty, M. and Clegg, T.L. (2019) The platformization of the classroom: Teachers as surveillant consumers. *Surveillance & Society*, 17: 145–152.

Kung, S., Fink, P., Hume, P. and Shultz, S. (2015) Kinematic and kinetic differences between barefoot and shod walking in children. *Footwear Science*, 7: 95–105.

Land, N. (2017) *Fat(s), muscle(s), movement, and physiologies in early childhood education* (Doctoral dissertation, University of Victoria, Canada).

Land, N., Hamm, C., Yazbeck, S.L., Danis, I., Brown, M. and Nelson, N. (2019, online early) Facetiming common worlds: Exchanging digital place stories and crafting pedagogical contact zones. *Children's Geographies*.

Latour, B. (2005) *Reassembling the social: An introduction to Actor-Network Theory*. Oxford: Oxford University Press.

Leck, H., Conway, D., Bradshaw, M. and Rees, J. (2015) Tracing the water-energy-food nexus: Description, theory and practice. *Geography Compass*, 9: 445–460.

Lee, N. and Motzkau, J. (2011) Navigating the bio-politics of childhood. *Childhood*, 18: 7–19.

Lenz-Taguchi, H. (2014) New materialisms and play. In *Sage handbook of play and learning in early childhood* (pp. 79–90). London: SAGE.

Leong, D. (2016) The mattering of Black lives: Octavia Butler's hyperempathy and the promise of the new materialisms. *Catalyst: Feminism, Theory, Technoscience*, 2: 1–35.

Levison, D., DeGraff, D. and Dungumaro, E. (2018) Implications of environmental chores for schooling: Children's time fetching water and firewood in Tanzania. *The European Journal of Development Research*, 30: 217–234.

Livingstone, S. (2019) Audiences in an age of datafication: Critical questions for media research. *Television & New Media*, 20: 170–183.

Livingstone, S., Lemish, D., Lim, S.S., Bulger, M., Cabello, P., Claro, M., Cabello-Hutt, T., Khalil, J., Kumpulainen, K., Nayar, U.S. and Nayar, P. (2017) Global perspectives on children's digital opportunities: An emerging research and policy agenda. *Pediatrics*, 140 (Supplement 2): S137–S141.

Livingstone, S. and Third, A. (2017) Children and young people's rights in the digital age: An emerging agenda. *New Media and Society*, 19: 657–670.

Loh, K.K. and Kanai, R. (2016) How has the Internet reshaped human cognition? *The Neuroscientist*, 22: 506–520.

Long, M., Moriceau, B., Gallinari, M., Lambert, C., Huvet, A., Raffray, J. and Soudant, P. (2015) Interactions between microplastics and phytoplankton aggregates: Impact on their respective fates. *Marine Chemistry*, 175: 39–46.

Lorimer, J. and Driessen, C. (2014) Wild experiments at the Oostvaardersplassen: Rethinking environmentalism in the Anthropocene. *Transactions of the Institute of British Geographers*, 39: 169–181.

Lulle, A. (2018) Relational ageing: On intra-gender and generational dynamism among ageing Latvian women. *Area*, 50: 452–458.

Lusinga, S. and de Groot, J. (2019) Energy consumption behaviours of children in low-income communities: A case study of Khayelitsha, South Africa. *Energy Research and Social Science*, 54: 199–210.

Maclean, M., Russell, W. and Ryall, E. (2015) *Philosophical perspectives on play*. London: Routledge.

Madanipour, A. (2018) Temporary use of space: Urban processes between flexibility, opportunity and precarity. *Urban Studies*, 55: 1110.

Madge, C. (2018) Creative geographies and living on from breast cancer: The enlivening potential of autobiographical bricolage for an aesthetics of precarity. *Transactions of the Institute of British Geographers*, 43: 245–261.

Malbon, B. (1999) *Clubbing: Clubbing culture and experience*. London: Routledge.

Mascheroni, G. (2018) Researching datafied children as data citizens. *Journal of Children and Media*, 12: 517–523.

Massey, D. (2005) *For space*. London: SAGE.

Matthews, H., Taylor, M., Percy-Smith, B. and Limb, M. (2000) The unacceptable flaneur: The shopping mall as a teenage hangout. *Childhood*, 7: 279–294.

McCormack, D.P. (2003) An event of geographical ethics in spaces of affect. *Transactions of the Institute of British Geographers*, 28: 488–507.

McCormack, D.P. (2017) The circumstances of post-phenomenological life worlds. *Transactions of the Institute of British Geographers*, 42: 2–13.

McKinnon, I., Hurley, P.T., Myles, C.C., Maccaroni, M. and Filan, T. (2019) Uneven urban metabolisms: Toward an integrative (ex) urban political ecology of sustainability in and around the city. *Urban Geography*, 40: 352–377.

Meillassoux, Q. (2010) *After finitude: An essay on the necessity of contingency*. London: Bloomsbury Publishing.

Merewether, J. (2019) New materialisms and children's outdoor environments: Murmurative diffractions. *Children's Geographies*, 17: 105–117.

Merritt, E., Bowers, N. and Rimm-Kaufman, S. (2019) Making connections: Elementary students' ideas about electricity and energy resources. *Renewable Energy*, 138: 1078–1086.

Miller, A.S. (2013) *Speculative grace: Bruno Latour and object-oriented theology*. Lincoln: Fordham University Press.

Mills, E. (2016) Identifying and reducing the health and safety impacts of fuel-based lighting. *Energy for Sustainable Development*, 30: 39–50.

Mitchell, K. and Elwood, S. (2012) Mapping children's politics: The promise of articulation and the limits of nonrepresentational theory. *Environment and Planning D: Society and Space*, 30: 788–804.

Morton, T. (2010) *The ecological thought*. Cambridge, MA: Harvard University Press.

Morton, T. (2013) *Hyperobjects: Philosophy and ecology after the end of the world*. Minneapolis: University of Minnesota Press.

Mumford, L. (1934) *Technics and civilization*. New York and Burlingame: Harcourt, Brace & Jovanovich.

Mumford, L. (1967) *The myth of the machine: Technics and human development* (Vol. 1). San Diego: Harcourt Brace Jovanovich.

Murris, K. (2016) *The posthuman child: Educational transformation through philosophy with picturebooks*. London: Routledge.

Mycock, K. (2019) Playing with mud-becoming stuck, becoming free? . . . The negotiation of gendered/class identities when learning outdoors. *Children's Geographies*, 17: 454–466.

Nail, T. (2017) What is an Assemblage? *SubStance*, 46: 21–37.

Natarajan, N., Parsons, L. and Brickell, K. (2019, online early) Debt-bonded Brick Kiln workers and their intent to return: Towards a labour geography of smallholder farming persistence in Cambodia. *Antipode*.

Nayak, A. (2010) Race, affect, and emotion: Young people, racism, and graffiti in the postcolonial English suburbs. *Environment and Planning A*, 42: 2370–2392.

Negarestani, R. (2008) *Cyclonopedia: Complicity with anonymous materials*. Melbourne: re.press.

Neimanis, A. (2017) *Bodies of water: Posthuman feminist phenomenology*. London: Bloomsbury Publishing.

Nolt, J. (2004) An argument for metaphysical realism. *Journal for General Philosophy of Science*, 35: 71–90.

Nordstrom, S.N. (2018) Antimethodology: Postqualitative generative conventions. *Qualitative Inquiry*, 24: 215–226.

Novotna, K., Cermakova, L., Pivokonska, L., Cajthaml, T. and Pivokonsky, M. (2019) Microplastics in drinking water treatment: Current knowledge and research needs. *Science of the Total Environment*, 667: 730–740.

Nxumalo, F. (2018) Situating indigenous and Black childhoods in the anthropocene. In *Research handbook on childhoodnature: Assemblages of childhood and nature research* (pp. 1–22). Berlin: Springer.

Nxumalo, F. and Cedillo, S. (2017) Decolonizing place in early childhood studies: Thinking with indigenous onto-epistemologies and Black feminist geographies. *Global Studies of Childhood*, 7: 99–112.

Osborne, T. and Jones, P.I. (2017) Biosensing and geography: A mixed methods approach. *Applied Geography*, 87: 160–169.

O'Sullivan, K.C., Howden-Chapman, P., Sim, D., Stanley, J., Rowan, R.L., Clark, I.K.H., Morrison, L.L. and Waiopehu College 2015 Research Team (2017) Cool? Young people investigate living in cold housing and fuel poverty: A mixed methods action research study. *SSM-Population Health*, 3: 66–74.

Paakkari, A., Rautio, P. and Valasmo, V. (2019) Digital labour in school: Smartphones and their consequences in classrooms. *Learning, Culture and Social Interaction*, 21: 161–169.

Pacini-Ketchabaw, V. and Clark, V. (2016) Following watery relations in early childhood pedagogies. *Journal of Early Childhood Research*, 14: 98–111.

Pacini-Ketchabaw, V. and Taylor, A. (2015) *Unsettling the colonial places and spaces of early childhood education*. London: Routledge.

Paddison, R., Philo, C., Routledge, P. and Sharp, J. (2000) *Entanglements of power: Geographies of domination/resistance*. London: Routledge.

Palmer, S. (2015) *Toxic childhood: How the modern world is damaging our children and what we can do about it*. London: Orion.

Parikka, J. (2013) *What is media archaeology?* London: John Wiley & Sons.

Patel, S., Patel, S. and Kumar, A. (2019) Effects of cooking fuel sources on the respiratory health of children: Evidence from the Annual Health Survey, Uttar Pradesh, India. *Public Health*, 169: 59–68.

Perrotta, C. and Williamson, B. (2018) The social life of learning analytics: Cluster analysis and the 'performance' of algorithmic education. *Learning, Media and Technology*, 43: 3–16.

Petersen, E.B. (2018) 'Data found us': A critique of some new materialist tropes in educational research. *Research in Education*, 101: 5–16.

Petrova, S. (2018) Encountering energy precarity: Geographies of fuel poverty among young adults in the UK. *Transactions of the Institute of British Geographers*, 43: 17–30.

Philo, C., Cadman, L. and Lea, J. (2015) New energy geographies: A case study of yoga, meditation and healthfulness. *Journal of Medical Humanities*, 36: 35–46.

Plowman, L. (2019) When the technology disappears. In *Exploring key issues in early childhood and technology: Evolving perspectives and innovative approaches* (pp. 32–37). Abingdon: Taylor & Francis.

Postman, N. (1985) The disappearance of childhood. *Childhood Education*, 61: 286–293.

Prout, A. (2005) *The future of childhood*. London: Routledge.

Punch, S. (2002) Youth transitions and interdependent adult-child relations in rural Bolivia. *Journal of Rural Studies*, 18: 123–133.

Punch, S. (2019, online early) Why have generational orderings been marginalised in the social sciences including childhood studies? *Children's Geographies*.

Purhonen, S. (2016) The modern meaning of the concept of generation. In *The Routledge international handbook on narrative and life history* (p. 167). London: Routledge.

Pyer, M., Horton, J., Tucker, F., Ryan, S. and Kraftl, P. (2010) Children, young people and 'disability': Challenging children's geographies? *Children's Geographies*, 8: 1–8.

Pykett, J. (2009) Personalization and de-schooling: Uncommon trajectories in contemporary education policy. *Critical Social Policy*, 29: 374–397.

Rautio, P. (2013) Children who carry stones in their pockets: On autotelic material practices in everyday life. *Children's Geographies*, 11: 394–408.

Rekret, P. (2016) A critique of new materialism: Ethics and ontology. *Subjectivity*, 9: 225–245.

Ringler, C., Bhaduri, A. and Lawford, R. (2013) The nexus across water, energy, land and food (WELF): Potential for improved resource use efficiency? *Current Opinion in Environmental Sustainability*, 5: 617–624.

Ringrose, J., Harvey, L., Gill, R. and Livingstone, S. (2013) Teen girls, sexual double standards and 'sexting': Gendered value in digital image exchange. *Feminist Theory*, 14: 323.

Robson, S.M., Couch, S.C., Peugh, J.L., Glanz, K., Zhou, C., Sallis, J.F. and Saelens, B.E. (2016) Parent diet quality and energy intake are related to child diet quality and energy intake. *Journal of the Academy of Nutrition and Dietetics*, 116: 984–990.

Rooney, T. (2019a) Weathering time: Walking with young children in a changing climate. *Children's Geographies*, 17: 177–189.

Rooney, T. (2019b) Sticking: Children and the lively matter of sticks. In *Feminist research for 21st-century childhoods: Common worlds methods* (p. 43). London: Bloomsbury.

Rose, G. (2016) Rethinking the geographies of cultural 'objects' through digital technologies: Interface, network and friction. *Progress in Human Geography*, 40: 334–351.

Rose, N. (2001) The politics of life itself. *Theory, Culture & Society*, 18: 1–30.

Rosen, R. (2015) 'The scream': Meanings and excesses in early childhood settings. *Childhood*, 22: 39–52.

Ryan, K.W. (2012) The new wave of childhood studies: Breaking the grip of bio-social dualism? *Childhood*, 19: 439–452.

Schatzki, T. (2016) Practice theory as flat ontology. In *Practice theory and research* (pp. 44–58). London: Routledge.

Schmidt, C., Krauth, T. and Wagner, S. (2017) Export of plastic debris by rivers into the sea. *Environmental Science and Technology*, 51: 12246–12253.

Schwanen, T. (2018) Thinking complex interconnections: Transition, nexus and geography. *Transactions of the Institute of British Geographers*, 43: 262–283.

Selwyn, N. (2003) Doing IT for the kids': Re-examining children, computers and the information society. *Media, Culture & Society*, 25: 351–378.

Sharma, R., Choudhary, D., Kumar, P., Venkateswaran, J. and Singh Solanki, C. (2019) Do solar study lamps help children study at night? Evidence from rural India. *Energy for Sustainable Development*, 50: 109–116.

Smith, T.A. and Dunkley, R. (2018) Technology-nonhuman-child assemblages: Reconceptualising rural childhood roaming. *Children's Geographies*, 16: 304–318.

Sothern, M. (2007) HIV+ bodyspace: AIDS and the queer politics of future negation in Aotearoa/New Zealand. In *Geographies of sexualities*. London: Routledge.

Sovacool, B.K. (2011) An international comparison of four polycentric approaches to climate and energy governance. *Energy Policy*, 39: 3832–3844.

Sovacool, B.K. (2014) What are we doing here? Analyzing fifteen years of energy scholarship and proposing a social science research agenda. *Energy Research & Social Science*, 1: 1–29.

Spill, M., Birch, L., Roe, L. and Rolls, B. (2011) Hiding vegetables to reduce energy density: An effective strategy to increase children's vegetable intake and reduce energy intake. *The American Journal of Clinical Nutrition*, 94: 735–741.

Spyrou, S. (2017) Time to decenter childhood? *Childhood*, 24: 433–437.

Stengers, I. (2015) *In catastrophic times: Resisting the coming barbarism*. London: Open Humanities Press.

Strengers, Y., Nicholls, L. and Maller, C. (2016) Curious energy consumers: Humans and nonhumans in assemblages of household practices. *Journal of Consumer Culture*, 16: 761–780.

Stryker, R. and Yngvesson, B. (2013) Fixity and fluidity: Circulations, children, and childhood. *Childhood*, 20: 297–306.

Swist, T., Collin, P. and Third, A. (2019) Children's data journeys beyond the 'supply chain': Co-curating flows and frictions. *Media International Australia*, 170: 68–77.

Taylor, A. (2013) Caterpillar childhoods: Engaging the otherwise worlds of Central Australian Aboriginal children. *Global Studies of Childhood*, 3: 366–379.

Taylor, A. (2019, online early) Countering the conceits of the Anthropos: Scaling down and researching with minor players. *Discourse: Studies in the Cultural Politics of Education*.

Taylor, A., Blaise, M. and Giugni, M. (2013) Haraway's 'bag lady story-telling': Relocating childhood and learning within a 'post-human landscape'. *Discourse: Studies in the Cultural Politics of Education*, 34: 48–62.

Taylor, A. and Pacini-Ketchabaw, V. (2018) *The common worlds of children and animals: Relational ethics for entangled lives*. London: Routledge.

Taylor, P.J., Evans, D.M. and Pain, K. (2008) Application of the interlocking network model to mega-city-regions: Measuring polycentricity within and beyond city-regions. *Regional Studies*, 42: 1079–1093.

Thompson, G. and Sellar, S. (2018) Datafication, testing events and the outside of thought. *Learning, Media and Technology*, 43: 139–151.

Thorne, B. (2007) Crafting the interdisciplinary field of childhood studies. *Childhood*, 14: 147–152.

Tian, X. (2017) Ethnobotanical knowledge acquisition during daily chores: The firewood collection of pastoral Maasai girls in Southern Kenya. *Journal of Ethnobiology and Ethnomedicine*, 13: unpaginated.

Toth, N., Little, L., Read, J., Fitton, D. and Horton, M. (2013) Understanding teen attitudes towards energy consumption. *Journal of Environmental Psychology*, 34: 36–44.

Trajber, R. and Mochizuki, Y. (2015) Climate change education for sustainability in Brazil: A status report. *Journal of Education for Sustainable Development*, 9: 44–61.

Trajber, R., Walker, C., Marchezini, V., Kraftl, P., Olivato, D., Hadfield-Hill, S., Zara, C. and Fernandes Monteiro, S. (2019) Promoting climate change transformation with young people in Brazil: Participatory action research through a looping approach. *Action Research*, 17: 87–107.

Tranter, P. and Sharpe, S. (2012) Pixar to the rescue: Harnessing positive affect for enhancing children's mobility. *Journal of Transport Geography*, 20: 34–40.

Tsing, A. (2010) Arts of inclusion, or how to love a mushroom. *Manoa*, 22: 191–203.

Tsing, A. (2015) *The mushroom at the end of the world: On the possibility of life in capitalist ruins*. Princeton: Princeton University Press.

Turkle, S. (2017) *Alone together: Why we expect more from technology and less from each other*. London: Hachette UK.

UNICEF (2014) *The challenges of climate change: Children on the front line*. New York: UNICEF.

UNICEF (2019) *#Everychild2030: Key asks and principles for 2018 national review activities*. New York: UNICEF.

Uprichard, E. (2008) Children as 'being and becomings': Children, childhood and temporality. *Children & Society*, 22: 303–313.

Valentine, G. (1996) Angels and devils: Moral landscapes of childhood. *Environment and Planning D: Society and Space*, 14: 581–599.

van Blerk, L. (2019) Where in the world are youth geographies going? Reflections on the journey and directions for the future. *Children's Geographies*, 17: 32–35.

Van Dijck, J., Poell, T. and De Waal, M. (2018) *The platform society: Public values in a connective world*. Oxford: Oxford University Press.

Von Uexküll, J. (2013) *A foray into the worlds of animals and humans: With a theory of meaning* (Vol. 12). Minneapolis: University of Minnesota Press.

Wachelder, J. (2019) Regeneration: Generations remediated. *Time & Society*, 28: 883–903.

Walker, C. (2019, online early) Nexus thinking and the geographies of children, youth and families: Towards an integrated research agenda. *Children's Geographies*.

Wallis, H., Nachreiner, M. and Matthies, E. (2016) Adolescents and electricity consumption: Investigating sociodemographic, economic, and behavioural influences on electricity consumption in households. *Energy Policy*, 94: 224–234.

Waltz, E. (2018) Therapy robot teaches social skills to children with autism. *IEEE Spectrum*, 9 August. Available online at: https://spectrum.ieee.org/the-human-os/biomedical/devices/robot-therapy-for-autism, last accessed 15th August 2019.

Ward, C. (1978) *The child in the city*. London: Pantheon.

Wells, K. (2011) The politics of life: Governing childhood. *Global Studies of Childhood*, 1: 15–25.

Wells, K. (2013) The melodrama of being a child: NGO representations of poverty. *Visual Communication*, 12: 277–293.

White, J. (2017) Climate change and the generational timescape. *The Sociological Review*, 65: 763–778.

WHO [World Health Organization] (2019) *Microplastics in drinking water*. Available online at: https://www.who.int/water_sanitation_health/publications/microplastics-in-drinking-water/en/, last accessed 20th January 2020.

Williamson, B. (2017) Learning in the 'platform society': Disassembling an educational data assemblage. *Research in Education*, 98: 59–82.

Willson, M. (2018, online early) Raising the ideal child? Algorithms, quantification and prediction. *Media, Culture & Society*.

Windram-Geddes, M. (2013) Fearing fatness and feeling fat: Encountering affective spaces of physical activity. *Emotion, Space and Society*, 9: 42–49.

Wolgemuth, J.R., Rautio, P., Koro-Ljungberg, M., Marn, T.M., Nordstrom, S. and Clark, A. (2018) Work/think/play/birth/death/terror/qualitative/research. *Qualitative Inquiry*, 24: 712–719.

Woodman, D. and Bennett, A. (2016) *Youth cultures, transitions, and generations: Bridging the gap in youth research*. Berlin: Springer.

Woodman, D. and Wyn, J. (2014) *Youth and generation: Rethinking change and inequality in the lives of young people*. London: SAGE.

Woodyer, T. (2012) Ludic geographies: Not merely child's play. *Geography Compass*, 6: 313–326.

World Bank (2011) *São Paulo Case Study Overview: Climate Change, Disaster Risk and the Urban Poor: Cities Building Resilience for a Changing World*. Available online at: http://siteresources.worldbank.org/INTURBANDEVELOPMENT/Resources/336387-1306291319853/CS_Sao_Paulo.pdf, last accessed 18th October 2019.

Worth, N. (2009) Understanding youth transition as 'becoming': Identity, time and futurity. *Geoforum*, 40: 1050–1060.

Worth, N. (2018) Mothers, daughters, and learning to labour: Framing work through gender and generation. *The Canadian Geographer/Le Géographe canadien*, 62: 551–561.

Wyness, M. (2013) Global standards and deficit childhoods: The contested meaning of children's participation. *Children's Geographies*, 11: 340–353.

Yamaguchi, Y., Soki, N. and Yoshiyuki, S. (2012) Per capita energy consumption for living, work, transport and other activities in the Keihanshin Metropolitan Region, Japan. *International Journal of Sustainable Building Technology and Urban Development*, 3: 68–76.

Youdell, D. (2017) Bioscience and the sociology of education: The case for biosocial education. *British Journal of Sociology of Education*, 38: 1273–1287.

Yusoff, K. (2017) Geosocial strata. *Theory, Culture & Society*, 34: 105–127.

Zalasiewicz, J., Waters, C.N., do Sul, J.A.I., Corcoran, P.L., Barnosky, A.D., Cearreta, A., Edgeworth, M., Gałuszka, A., Jeandel, C., Leinfelder, R. and McNeill, J.R. (2016) The geological cycle of plastics and their use as a stratigraphic indicator of the Anthropocene. *Anthropocene*, 13: 4–17.

Zizek, S. (2006) *The parallax view*. Cambridge, MA: The MIT Press.

Index

Note: Numbers in italic indicate a figure and page numbers in bold indicate a table on the corresponding page.

Printed in the United States
by Baker & Taylor Publisher Services

Printed in the United States
by Baker & Taylor Publisher Services